Analysis of
Generalized Linear Mixed Models
in the Agricultural and Natural Resources Sciences

Analysis of
Generalized Linear Mixed Models
in the Agricultural and Natural Resources Sciences

**Edward E. Gbur, Walter W. Stroup,
Kevin S. McCarter, Susan Durham,
Linda J. Young, Mary Christman,
Mark West, and Matthew Kramer**

American Society
of Agronomy

Soil
Science
Society of America

Crop Science
Society of America

American Society of Agronomy
Soil Science Society of America
Crop Science Society of America, Inc.
5585 Guilford Road, Madison, WI 53711-5801 USA
https://www.agronomy.org/publications/books I www.SocietyStore.org

ISBN: 978-0-89118-182-8
e-ISBN: 978-0-89118-183-5
doi:10.2134/2012.generalized-linear-mixed-models

Library of Congress Control Number: 2011944082

Cover: Patricia Scullion
Photo: Nathan Slaton, Univ. of Arkansas, Dep. of Crops, Soil, and Environmental Science

Printed in the United States of America.

CONTENTS

Analysis of Generalized Linear Mixed Models in the Agricultural and Natural Resources Sciences is an excellent resource book for students and professionals alike. This book explains the use of generalized linear mixed models which are applicable to students of agricultural and natural resource sciences. The strength of the book is the available examples and statistical analysis system (SAS) code used for analysis. These "real life" examples provide the reader with the examples needed to understand and use generalized linear mixed models for their own analysis of experimental data. This book, published by the American Society of Agronomy, Crop Science Society of America, and the Soil Science Society of America, will be valuable as its practical nature will help scientists in training as well as practicing scientists. The goal of the three Societies is to provide educational material to advance the profession. This book helps meet this goal.

Chuck Rice, 2011 Soil Science Society of America President
Newell Kitchen, 2011 American Society of Agronomy President
Maria Gallo, 2011 Crop Science Society of America President

PREFACE

The authors of this book are participants in the Multi-state Project NCCC-170 "Research Advances in Agricultural Statistics" under the auspices of the North Central Region Agricultural Experiment Station Directors. Project members are statisticians from land grant universities, USDA-ARS, and industry who are interested in agricultural and natural resource applications of statistics. The project has been in existence since 1991. We consider this book as part of the educational outreach activities of our group. Readers interested in NCCC-170 activities can access the project website through a link on the National Information Management and Support System (NIMSS).

Traditional statistical methods have been developed primarily for normally distributed data. Generalized linear mixed models extend normal theory linear mixed models to include a broad class of distributions, including those commonly used for counts, proportions, and skewed distributions. With the advent of software for implementing generalized linear mixed models, we have found researchers increasingly interested in using these models, but it is "easier said than done." Our goal is to help those who have worked with linear mixed models to begin moving toward generalized linear mixed models. The benefits and challenges are discussed from a practitioner's viewpoint. Although some readers will feel confident in fitting these models after having worked through the examples, most will probably use this book to become aware of the potential these models promise and then work with a professional statistician for full implementation, at least for their first few applications.

The original purpose of this book was as an educational outreach effort to the agricultural and natural resources research community. This remains as its primary purpose, but in the process of preparing this work, each of us found it to be a wonderful professional development experience. Each of the authors understood some aspects of generalized linear mixed models well, but no one "knew it all." By pooling our combined understanding and discussing different perspectives, we each have benefitted greatly. As a consequence, those with whom we consult will benefit from this work as well.

We wish to thank our reviewers Bruce Craig, Michael Guttery, and Margaret Nemeth for their careful reviews and many helpful comments. Jeff Velie constructed many of the graphs that were not automatically generated by SAS (SAS Institute, Cary, NC). Thank you, Jeff. We are grateful to all of the scientists who so willingly and graciously shared their research data with us for use as examples.

Edward E. Gbur, Walter W. Stroup, Kevin S. McCarter, Susan Durham,
Linda J. Young, Mary Christman, Mark West, and Matthew Kramer

Edward Gbur is currently Professor and Director of the Agricultural Statistics Laboratory at the University of Arkansas. Previously he was on the faculty in the Statistics Department at Texas A&M University and was a Mathematical Statistician in the Statistical Research Division at the Census Bureau. He received a Ph.D. in Statistics from The Ohio State University. He is a member and Fellow of the American Statistical Association and a member of the International Biometric Society and the Institute of Mathematical Statistics. His current research interests include experimental design, generalized linear mixed models, stochastic modeling, and agricultural applications of statistics.

Walter Stroup is Professor of Statistics at the University of Nebraska, Lincoln. After receiving his Ph.D. in Statistics from the University of Kentucky in 1979, he joined the Biometry faculty at Nebraska's Institute of Agriculture and Natural Resources. He served as teacher, researcher, and consultant until becoming department chair in 2001. In 2003, Biometry was incorporated into a new Department of Statistics at UNL; Walt served as chair from its founding through 2010. He is co-author of *SAS for Mixed Models* and *SAS for Linear Models*. He is a member of the International Biometric Society, American Association for the Advancement of Science, and a member and Fellow of the American Statistical Association. His interests include design of experiments and statistical modeling.

Kevin S. McCarter is a faculty member in the Department of Experimental Statistics at Louisiana State University. He earned the Bachelors degree with majors in Mathematics and Computer Information Systems from Washburn University and the Masters and Ph.D. degrees in Statistics from Kansas State University. He has industry experience as an IT professional in banking, accounting, and health care, and as a biostatistician in the pharmaceutical industry. His dissertation research was in the area of survival analysis. His current research interests include predictive modeling, developing and assessing statistical methodology, and applying generalized linear mixed modeling techniques. He has collaborated with researchers from a wide variety of fields, including agriculture, biology, education, medicine, and psychology.

Susan Durham is a statistical consultant at Utah State University, collaborating with faculty and graduate students in the Ecology Center, Biology Department, and College of Natural Resources. She earned a Bachelors degree in Zoology at Oklahoma State University and a Masters degree in Applied Statistics at Utah State University. Her interests cover the broad range of research problems that have been brought to her as a statistical consultant.

Mary Christman is currently the lead statistical consultant with MCC Statistical Consulting LLC, which provides statistical expertise for environmental and ecological problems. She is also courtesy professor at the University of Florida. She was on the faculty at University of Florida, University of Maryland, and American University after receiving her Ph.D. in statistics from George Washington University. She is a member of several organizations, including the American Statistical Association, the International Environmetrics Society, and the American Association for the Advancement of Science. She received the 2004 Distinguished Achievement Award from the Section on Statistics and the Environment of the American Statistical Association. Her current research interests include linear and non-linear modeling in the presence of correlated error terms, sampling and experimental design, and statistical methodology for ecological and environmental research.

Linda J. Young is Professor of Statistics at the University of Florida. She completed her Ph.D. in Statistics at Oklahoma State University and has previously served on the faculties of Oklahoma State University and the University of Nebraska, Lincoln. Linda has served the profession in a variety of capacities, including President of the Eastern North American Region of the International Biometric Society, Treasurer of the International Biometric Society, Vice-President of the American Statistical Association, and Chair of the Committee of Presidents of Statistical Societies. She has co-authored two books and has more than 100 refereed publications. She is a fellow of the American Association for the Advancement of Science, a fellow of the American Statistical Association, and an elected member of the International Statistical Institute. Her research interests include spatial statistics and statistical modeling.

Mark West is a statistician for the USDA-Agricultural Research Service. He received his Ph.D. in Applied Statistics from the University of Alabama in 1989 and has been a statistical consultant in agriculture research ever since beginning his professional career at Auburn University in 1989. His interests include experimental design, statistical computing, computer intensive methods, and generalized linear mixed models.

Matt Kramer is a statistician in the mid-Atlantic area (Beltsville, MD) of the USDA-Agricultural Research Service, where he has worked since 1999. Prior to that, he spent eight years at the Census Bureau in the Statistical Research Division (time series and small area estimation). He received a Masters and Ph.D. from the University of Tennessee. His interests are in basic biological and ecological statistical applications.

To convert Column 1 into Column 2 multiply by	Column 1 SI unit	Column 2 non-SI unit	To convert Column 2 into Column 1 multiply by
Length			
0.621	kilometer, km (10^3 m)	mile, mi	1.609
1.094	meter, m	yard, yd	0.914
3.28	meter, m	foot, ft	0.304
1.0	micrometer, µm (10^{-6} m)	micron, µ	1.0
3.94×10^{-2}	millimeter, mm (10^{-3} m)	inch, in	25.4
10	nanometer, nm (10^{-9} m)	Angstrom, Å	0.1
Area			
2.47	hectare, ha	acre	0.405
247	square kilometer, km^2 (10^3 m)^2	acre	4.05×10^{-3}
0.386	square kilometer, km^2 (10^3 m)^2	square mile, mi^2	2.590
2.47×10^{-4}	square meter, m^2	acre	4.05×10^3
10.76	square meter, m^2	square foot, ft^2	9.29×10^{-2}
1.55×10^{-3}	square millimeter, mm^2 (10^{-3} m)^2	square inch, in^2	645
Volume			
9.73×10^{-3}	cubic meter, m^3	acre-inch	102.8
35.3	cubic meter, m^3	cubic foot, ft^3	2.83×10^{-2}
6.10×10^4	cubic meter, m^3	cubic inch, in^3	1.64×10^{-5}
2.84×10^{-2}	liter, L (10^{-3} m^3)	bushel, bu	35.24
1.057	liter, L (10^{-3} m^3)	quart (liquid), qt	0.946
3.53×10^{-2}	liter, L (10^{-3} m^3)	cubic foot, ft^3	28.3
0.265	liter, L (10^{-3} m^3)	gallon	3.78
33.78	liter, L (10^{-3} m^3)	ounce (fluid), oz	2.96×10^{-2}
2.11	liter, L (10^{-3} m^3)	pint (fluid), pt	0.473
Mass			
2.20×10^{-3}	gram, g (10^{-3} kg)	pound, lb	454
3.52×10^{-2}	gram, g (10^{-3} kg)	ounce (avdp), oz	28.4
2.205	kilogram, kg	pound, lb	0.454
0.01	kilogram, kg	quintal (metric), q	100
1.10×10^{-3}	kilogram, kg	ton (2000 lb), ton	907
1.102	megagram, Mg (tonne)	ton (U.S.), ton	0.907
1.102	tonne, t	ton (U.S.), ton	0.907
Yield and Rate			
0.893	kilogram per hectare, kg ha^{-1}	pound per acre, lb acre^{-1}	1.12
7.77×10^{-2}	kilogram per cubic meter, kg m^{-3}	pound per bushel, lb bu^{-1}	12.87
1.49×10^{-2}	kilogram per hectare, kg ha^{-1}	bushel per acre, 60 lb	67.19
1.59×10^{-2}	kilogram per hectare, kg ha^{-1}	bushel per acre, 56 lb	62.71

continued

To convert Column 1 into Column 2 multiply by	Column 1 SI unit	Column 2 non-SI unit	To convert Column 2 into Column 1 multiply by
1.86×10^{-2}	kilogram per hectare, kg ha^{-1}	bushel per acre, 48 lb	53.75
0.107	liter per hectare, L ha^{-1}	gallon per acre	9.35
893	tonne per hectare, t ha^{-1}	pound per acre, lb acre^{-1}	1.12×10^{-3}
893	megagram per hectare, Mg ha^{-1}	pound per acre, lb acre^{-1}	1.12×10^{-3}
0.446	megagram per hectare, Mg ha^{-1}	ton (2000 lb) per acre, ton acre^{-1}	2.24
2.24	meter per second, m s^{-1}	mile per hour	0.447

Specific Surface

10	square meter per kilogram, m^2 kg^{-1}	square centimeter per gram, cm^2 g^{-1}	0.1
1000	square meter per kilogram, m^2 kg^{-1}	square millimeter per gram, mm^2 g^{-1}	0.001

Density

1.00	megagram per cubic meter, Mg m^{-3}	gram per cubic centimeter, g cm^{-3}	1.00

Pressure

9.90	megapascal, MPa (10^6 Pa)	atmosphere	0.101
10	megapascal, MPa (10^6 Pa)	bar	0.1
2.09×10^{-2}	pascal, Pa	pound per square foot, lb ft^{-2}	47.9
1.45×10^{-4}	pascal, Pa	pound per square inch, lb in^{-2}	6.90×10^3

Temperature

1.00 (K – 273)	kelvin, K	Celsius, °C	1.00 (°C + 273)
(9/5 °C) + 32	Celsius, °C	Fahrenheit, °F	5/9 (°F – 32)

Energy, Work, Quantity of Heat

9.52×10^{-4}	joule, J	British thermal unit, Btu	1.05×10^3
0.239	joule, J	calorie, cal	4.19
10^7	joule, J	erg	10^{-7}
0.735	joule, J	foot-pound	1.36
2.387×10^{-5}	joule per square meter, J m^{-2}	calorie per square centimeter (langley)	4.19×10^4
10^5	newton, N	dyne	10^{-5}
1.43×10^{-3}	watt per square meter, W m^{-2}	calorie per square centimeter minute (irradiance), cal cm^{-2} min^{-1}	698

Transpiration and Photosynthesis

3.60×10^{-2}	milligram per square meter second, mg m^{-2} s^{-1}	gram per square decimeter hour, g dm^{-2} h^{-1}	27.8
5.56×10^{-3}	milligram (H$_2$O) per square meter second, mg m^{-2} s^{-1}	micromole (H$_2$O) per square centimeter second, µmol cm^{-2} s^{-1}	180
10^{-4}	milligram per square meter second, mg m^{-2} s^{-1}	milligram per square centimeter second, mg cm^{-2} s^{-1}	10^4
35.97	milligram per square meter second, mg m^{-2} s^{-1}	milligram per square decimeter hour, mg dm^{-2} h^{-1}	2.78×10^{-2}

continued

To convert Column 1 into Column 2 multiply by	Column 1 SI unit	Column 2 non-SI unit	To convert Column 2 into Column 1 multiply by
Plane Angle			
57.3	radian, rad	degrees (angle), °	1.75×10^{-2}
Electrical Conductivity, Electricity, and Magnetism			
10	siemen per meter, S m^{-1}	millimho per centimeter, mmho cm^{-1}	0.1
10^4	tesla, T	gauss, G	10^{-4}
Water Measurement			
9.73×10^{-3}	cubic meter, m^3	acre-inch, acre-in	102.8
9.81×10^{-3}	cubic meter per hour, m^3 h^{-1}	cubic foot per second, ft^3 s^{-1}	101.9
4.40	cubic meter per hour, m^3 h$^-$1	U.S. gallon per minute, gal min^{-1}	0.227
8.11	hectare meter, ha m	acre-foot, acre-ft	0.123
97.28	hectare meter, ha m	acre-inch, acre-in	1.03×10^{-2}
8.1×10^{-2}	hectare centimeter, ha cm	acre-foot, acre-ft	12.33
Concentration			
1	centimole per kilogram, cmol kg^{-1}	milliequivalent per 100 grams, meq 100 g^{-1}	1
0.1	gram per kilogram, g kg^{-1}	percent, %	10
1	milligram per kilogram, mg kg^{-1}	parts per million, ppm	1
Radioactivity			
2.7×10^{-11}	becquerel, Bq	curie, Ci	3.7×10^{10}
2.7×10^{-2}	becquerel per kilogram, Bq kg^{-1}	picocurie per gram, pCi g^{-1}	37
100	gray, Gy (absorbed dose)	rad, rd	0.01
100	sievert, Sv (equivalent dose)	rem (roentgen equivalent man)	0.01
Plant Nutrient Conversion			
	Elemental	Oxide	
2.29	P	P$_2$O$_5$	0.437
1.20	K	K$_2$O	0.830
1.39	Ca	CaO	0.715
1.66	Mg	MgO	0.602

CHAPTER 1

INTRODUCTION

1.1 INTRODUCTION

Over the past generation, dramatic advances have occurred in statistical methodology, many of which are relevant to research in the agricultural and natural resources sciences. These include more theoretically sound approaches to the analysis of spatial data; data taken over time; data involving discrete, categorical, or continuous but non-normal response variables; multi-location and/or multi-year data; complex split-plot and repeated measures data; and genomic data such as data from microarray and quantitative genetics studies. The development of generalized linear mixed models has brought together these apparently disparate problems under a coherent, unified theory. The development of increasingly user friendly statistical software has made the application of this methodology accessible to applied researchers.

The accessibility of generalized linear mixed model software has coincided with a time of change in the research community. Research budgets have been tightening for several years, and there is every reason to expect this trend to continue for the foreseeable future. The focus of research in the agricultural sciences has been shifting as the nation and the world face new problems motivated by the need for clean and renewable energy, management of limited natural resources, environmental stress, the need for crop diversification, the advent of precision agriculture, safety dilemmas, and the need for risk assessment associated with issues such as genetically modified crops. New technologies for obtaining data offer new and important possibilities but often are not suited for design and analysis using conventional approaches developed decades ago. With this rapid development comes the lack of accepted guidelines for how such data should be handled.

Researchers need more efficient ways to conduct research to obtain useable information with the limited budgets they have. At the same time, they need ways to meaningfully analyze and understand response variables that are very different from those covered in "traditional" statistical methodology. Generalized linear mixed models allow more versatile and informative analysis in these situations and, in the process, provide the tools to facilitate experimental designs tailored to

doi:10.2134/2012.generalized-linear-mixed-models.c1

Analysis of Generalized Linear Mixed Models in the Agricultural and Natural Resources Sciences
Edward E. Gbur, Walter W. Stroup, Kevin S. McCarter, Susan Durham, Linda J. Young, Mary Christman, Mark West, and Matthew Kramer

the needs of particular studies. Such designs are often quite different from conventional experimental designs. Thus, generalized linear mixed models provide an opportunity for a comprehensive rethinking of statistical practice in agricultural and natural resources research. This book provides a practical introductory guide to this topic.

1.2 GENERALIZED LINEAR MIXED MODELS

In introductory statistical methods courses taken by nearly every aspiring agricultural scientist in graduate school, statistical analysis is presented in some way, shape, or form as an attempt to make inferences on observations that are the sum of "explanatory" components and "random" components. In designed experiments and quasi-experiments (i.e., studies structured as closely as possible to designed experiments), "explanatory" means treatment effect and "random" means residual or random error. Thus, the formula

observed response = explanatory + random

expresses the basic building blocks of statistical methodology. This simple breakdown is necessarily elaborated into

observed response = treatment + design effects + error

where design effects include blocks and covariates. The observed response is inevitably interpreted as having a normal distribution and analysis of variance (ANOVA), regression, and analysis of covariance are presented as the primary methods of analysis. In contemporary statistics, such models are collectively referred to as linear models. In simple cases, a binomial distribution is considered for the response variable leading to logit analysis and logistic regression. Occasionally probit analysis is considered as well.

In contrast, consider what the contemporary researcher actually faces. Table 1–1 shows the types of observed response variables and explanatory model components that researchers are likely to encounter. Note that "conventional" statistical methodology taught in introductory statistics courses and widely considered as "standard statistical analysis" in agricultural research and journal publication is confined to the first row and occasionally the second row in the table. Obviously, the range of methods considered "standard" is woefully inadequate given the range of possibilities now faced by contemporary researchers.

This inadequacy has a threefold impact on potential advances in agricultural and applied research. First, it limits the types of analyses that researchers (and journal editors) will consider, resulting in cases where "standard methods" are a mismatch between the observed response and an explanatory model. Second, it limits researchers' imaginations when planning studies, for example through a lack of awareness of alternative types of response variables that contemporary statistical methods can handle. Finally, it limits the efficiency of experiments in that traditional designs, while optimized for normal distribution based ANOVA

TABLE 1-1. Statistical model scenarios corresponding to combinations of types of observed responses and explanatory model components.

		Explanatory model components			
		Fixed effects			
Type of response variable	Examples of distributions	Categorical	Continuous	Random effects	Correlated errors
Continuous, unbounded values, symmetric	normal	ANOVA†,‡,§,¶	regression †,‡,§,¶	split plot ANOVA‡,¶	−‡,¶
Categorical	binomial, multinomial	logit analysis§,¶	logistic regression §,¶	−¶	−¶
Count	Poisson, negative binomial	log-linear model §,¶	Poisson regression §,¶	−¶	−¶
Continuous, non-negative values	lognormal, gamma, beta	−§,¶	−§,¶	−¶	−¶
Time to event	exponential, gamma, geometric	−§,¶	−§,¶	−¶	−¶

† Linear model scenarios are limited to the first two cells in the first row of the table.

‡ Linear mixed model scenarios are limited to first row of the table.

§ Generalized linear model scenarios are limited to first two columns of the table.

¶ Generalized linear mixed model scenarios cover all cells shown in the table.

and regression, often are not well suited to the majority of the response variable–explanatory model combinations in Table 1–1.

Two major advances in statistical theory and methodology that occurred in the last half of the 20th century were the development of linear mixed models and generalized linear models. Mixed models incorporate random effects and correlated errors; that is, they deal with all four columns of explanatory model components in Table 1–1. Generalized linear models accommodate a large class of probability distributions of the response; that is, they deal with the response variable column in the table. The combination of mixed and generalized linear models, namely *generalized linear mixed models*, addresses the entire range of options for the response variable and explanatory model components (i.e., with all 20 combinations in Table 1–1). Generalized linear mixed models represent the primary focus of this book.

1.3 HISTORICAL DEVELOPMENT

Seal (1967) traced the origin of fixed effects models back to the development of least squares by Legendre in 1806 and Gauss in 1809, both in the context of problems in astronomy. It is less well known that the origin of random effects models can be ascribed to astronomy problems as well. Scheffé (1956) attributed early use of random effects to Airy in an 1861 publication. It was not until nearly 60 years later that Fisher (1918) formally introduced the terms *variance* and *analysis of variance* and utilized random effects models.

Fisher's 1935 first edition of *The Design of Experiments* implicitly discusses mixed models (Fisher, 1935). Scheffé (1956) attributed the first explicit expression of a mixed model equation to Jackson (1939). Yates (1940) developed methods to recover inter-block information in block designs that are equivalent to mixed model analysis with random blocks. Eisenhart (1947) formally identified random, fixed, and mixed models. Henderson (1953) was the first to explicitly use mixed model methodology for animal genetics studies. Harville (1976, 1977) published the formal overall theory of mixed models.

Although analyses of special cases of non-normally distributed responses such as probit analysis (Bliss, 1935) and logit analysis (Berkson, 1944) existed in the context of bioassays, standard statistical methods textbooks such as Steel et al. (1997) and Snedecor and Cochran (1989) dealt with the general problem of non-normality through the use of transformations. The ultimate purpose of transformations such as the logarithm, arcsine, and square root was to enable the researcher to obtain approximate analyses using the standard normal theory methods. Box and Cox (1964) proposed a general class of transformations that include the above as special cases. They too have been applied to allow use of normal theory methods.

Nelder and Wedderburn (1972) articulated a comprehensive theory of linear models with non-normally distributed response variables. They assumed that the response distribution belonged to the exponential family. This family of probability distributions contains a diverse set of discrete and continuous distributions, including all of those listed in Table 1–1. The models were referred to as generalized linear models (not to be confused with general linear models which has been used in reference to normally distributed responses only). Using the concept of quasi-likelihood, Wedderburn (1974) extended applicability of generalized linear models to certain situations where the distribution cannot be specified exactly. In these cases, if the observations are independent or uncorrelated and the form of the mean/variance ratio can be specified, it is possible to fit the model and obtain results similar to those which would have been obtained if the distribution had been known. The monograph by McCullagh and Nelder (1989) brought generalized linear models to the attention of the broader statistical community and with it, the beginning of research on the addition of random effects to these models—the development of generalized linear mixed models.

By 1992 the conceptual development of linear models through and including generalized linear mixed models had been accomplished, but the computational capabilities lagged. The first usable software for generalized linear models appeared in the mid 1980s, the first software for linear mixed models in the 1990s, and the first truly usable software for generalized linear mixed models appeared in the mid 2000s. Typically there is a 5- to 10-year lag between the introduction of the software and the complete appreciation of the practical aspects of data analyses using these models.

1.4 OBJECTIVES OF THIS BOOK

Our purpose in writing this book is to lead practitioners gently through the basic concepts and currently available methods needed to analyze data that can be modeled as a generalized linear mixed model. These concepts and methods require a change in mindset from normal theory linear models that will be elaborated on at various points in the following chapters. As with all new methodology, there is a learning curve associated with this material and it is important that the theory be understood at least at some intuitive level. We assume that the reader is familiar with the corresponding standard techniques for normally distributed responses and has some experience using these methods with statistical software such as SAS (SAS Institute, Cary, NC) or R (CRAN, www.r-project.org [verified 27 Sept. 2011]). While it is necessary to use matrix language in some places, we have attempted to keep the mathematical level as accessible as possible for the reader. We believe that readers who find the mathematics too difficult will still find much of this book useful. Numerical examples have been included throughout to illustrate the concepts. The emphasis in these examples is on illustration of the methodology and not on subject matter results.

Chapter 2 presents background on the exponential family of probability distributions and the likelihood based statistical inference methods used in the analysis of generalized linear mixed models. Chapter 3 introduces generalized linear models containing only fixed effects. Random effects and the corresponding mixed models having normally distributed responses are the subjects of Chapter 4. Chapter 5 begins the discussion of generalized linear mixed models. In Chapter 6, detailed analyses of two more complex examples are presented. Finally we turn to design issues in Chapter 7, where our purpose is to provide examples of a methodology that allows the researcher to plan studies involving generalized linear mixed models that directly address his/her primary objectives efficiently. Chapter 8 contains final remarks.

This book represents a first effort to describe the analysis of generalized linear mixed models in the context of applications in the agricultural sciences. We are still in that early period following the introduction of software capable of fitting these models, and there are some unresolved issues concerning various aspects of working with these methods. As examples are introduced in the following chapters, we will note some of the issues that a data analyst is likely to encounter and will provide advice as to the best current thoughts on how to handle them. One recurring theme that readers will notice, especially in Chapter 5, is that computing software defaults often must be overridden. With increased capability comes increased complexity. It is unrealistic to expect one-size-fits-all defaults for generalized linear mixed model software. As these situations arise in this book, we will explain what to do and why. The benefit for the additional effort is more accurate analysis and higher quality information per research dollar.

REFERENCES CITED

Berkson, J. 1944. Application of the logistic function to bio-assay. J. Am. Stat. Assoc. 39:357–365. doi:10.2307/2280041

Bliss, C.A. 1935. The calculation of the dose-mortality curve. Ann. Appl. Biol. 22:134–167. doi:10.1111/j.1744-7348.1935.tb07713.x

Box, G.E.P., and D.R. Cox. 1964. An analysis of transformations. J. R. Stat. Soc. Ser. B (Methodological) 26:211–252.

Eisenhart, C. 1947. The assumptions underlying the analysis of variance. Biometrics 3:1–21. doi:10.2307/3001534

Fisher, R.A. 1918. The correlation between relatives on the supposition of Mendelian inheritance. Trans. R. Soc. Edinb. 52:399–433.

Fisher, R.A. 1935. The design of experiments. Oliver and Boyd, Edinburgh.

Harville, D.A. 1976. Confidence intervals and sets for linear combinations of fixed and random effects. Biometrics 32:403–407. doi:10.2307/2529507

Harville, D.A. 1977. Maximum likelihood approaches to variance component estimation and to related problems. J. Am. Stat. Assoc. 72:320–338. doi:10.2307/2286796

Henderson, C.R. 1953. Estimation of variance and covariance components. Biometrics 9:226–252. doi:10.2307/3001853

Jackson, R.W.B. 1939. The reliability of mental tests. Br. J. Psychol. 29:267–287.

McCullagh, P., and J.A. Nelder. 1989. Generalized linear models. 2nd ed. Chapman and Hall, New York.

Nelder, J.A., and R.W.M. Wedderburn. 1972. Generalized linear models. J. R. Stat. Soc. Ser. A (General) 135:370–384. doi:10.2307/2344614

Scheffé, H. 1956. Alternative models for the analysis of variance. Ann. Math. Stat. 27:251–271. doi:10.1214/aoms/1177728258

Seal, H.L. 1967. The historical development of the Gauss linear model. Biometrika 54:1–24.

Snedecor, G.W., and W.G. Cochran. 1989. Statistical methods. 8th ed. Iowa State Univ. Press, Ames, IA.

Steel, R.G.D., J.H. Torrie, and D.A. Dickey. 1997. Principles and procedures of statistics: A biometrical approach. 3rd ed. McGraw-Hill, New York.

Wedderburn, R.W.M. 1974. Quasi-likelihood functions, generalized linear models and the Gauss-Newton method. Biometrika 61:439–447.

Yates, F. 1940. The recovery of interblock information in balanced incomplete block designs. Ann. Eugen. 10:317–325. doi:10.1111/j.1469-1809.1940.tb02257.x

BACKGROUND

2.1 INTRODUCTION

This chapter provides background material necessary for an understanding of generalized linear mixed models. It includes a description of the exponential family of probability distributions and several other commonly used distributions in generalized linear models. An important characteristic that distinguishes a non-normal distribution in this family from the normal distribution is that its variance is a function of its mean. As a consequence, these models have heteroscedastic variance structures because the variance changes as the mean changes. A familiar example of this is the binomial distribution based on n independent trials, each having success probability π. The mean is $\mu = n\pi$, and the variance is $n\pi(1 - \pi) = \mu(1 - \mu/n)$.

The method of least squares has been commonly used as the basis for estimation and statistical inference in linear models where the response is normally distributed. As an estimation method, least squares is a mathematical method for minimizing the sum of squared errors that does not depend on the probability distribution of the response. While suitable for fixed effects models with normally distributed data, least squares does not generalize well to models with random effects, non-normal data, or both. Likelihood based procedures provide an alternative approach that incorporates the probability distribution of the response into parameter estimation as well as inference. Inference for mixed and generalized linear models is based on a likelihood approach described in Sections 2.4 through 2.7.

The basic concepts of fixed and random effects and the formulation of mixed models are reviewed in Sections 2.8 through 2.10. The final section of this chapter discusses available software.

2.2 DISTRIBUTIONS USED IN GENERALIZED LINEAR MODELING

Probability distributions that can be written in the form

$$f(y \mid v, \phi) = \exp\left[\frac{t(y)\eta(v) - A(v)}{a(\phi)} + h(y, \phi)\right]$$

doi:10.2134/2012.generalized-linear-mixed-models.c2

Analysis of Generalized Linear Mixed Models in the Agricultural and Natural Resources Sciences
Edward E. Gbur, Walter W. Stroup, Kevin S. McCarter, Susan Durham, Linda J. Young, Mary Christman, Mark West, and Matthew Kramer

are said to be members of the exponential family of distributions. The function $f(y \mid v, \phi)$ is the probability distribution of the response variable Y given v and ϕ, the location and scale parameters, respectively. The functions $t(\cdot)$, $\eta(\cdot)$, $A(\cdot)$, $a(\cdot)$, and $h(\cdot)$ depend on either the data, the parameters or both as indicated. The quantity $\eta(v)$ is known as the natural parameter or canonical form of the parameter. As will be seen in Chapter 3, the canonical parameter $\theta = \eta(v)$ plays an important role in generalized linear models. The mean and variance of the random variable Y can be shown to be a function of the parameter v and hence, of θ. As a result, for members of the one parameter exponential family, the probability distribution of Y determines both the canonical form of the parameter and the form of the variance as a function of v.

EXAMPLE 2.1

The binomial distribution is usually written as

$$f\left(y \mid \pi \right) = P\left(Y = y \mid \pi \right) = \binom{n}{y} \pi^{y} \left(1 - \pi \right)^{n-y}$$

where $y = 0, \ldots, n$. Assuming that n is known, the distribution has one parameter π ($= v$). Rewriting this probability in exponential family form, we have

$$f\left(y \mid \pi \right) = \exp\left\{ \log\left[f\left(y \mid \pi \right) \right] \right\} = \exp\left[y \log\left(\frac{\pi}{1-\pi} \right) + n \log\left(1 - \pi \right) + \log\binom{n}{y} \right]$$

where we identify $t(y) = y$, $\eta\left(\pi \right) = \log\left(\frac{\pi}{1-\pi} \right)$, $A(\pi) = -n \log(1 - \pi)$, and

$h\left(y \right) = \log\binom{n}{y}$ in the general form. Here log is the natural logarithm. For the bino-

mial distribution, $\phi = 1$, so that $a(\phi) = 1$. The canonical parameter

$$\theta = \eta\left(\pi \right) = \log\left(\frac{\pi}{1-\pi} \right)$$

is often referred to as the logit of π. ∎

The scale parameter ϕ is either a fixed and known positive constant (usually 1) or a parameter that must be estimated. Except for the normal distribution, the scale parameter does not correspond to the variance of Y. When ϕ is known, the family is referred to as a one parameter exponential family. An example of a one-parameter exponential family is the binomial distribution with parameters n, the sample size or number of trials, and π, the probability of a success. In this case, $\phi = 1$, and n is typically known. When ϕ is unknown, the family is referred to as a two parameter exponential family. An example of a two parameter exponential family is the normal distribution where, for generalized linear model purposes, v is the mean and ϕ is the variance. Another example of a two parameter distribution is the gamma distribution where v is the mean and ϕv^2 is the variance. Note that for the normal distribution the mean and variance are distinct parameters, but for the gamma distribution the variance depends on both the mean and the scale parameters. Other distributions in which the variance depends on both the mean and scale parameters include the beta and negative binomial distributions (Section 2.3).

EXAMPLE 2.2

The normal distribution with mean μ and variance σ^2 is usually written as

$$f\left(y \mid \mu, \sigma^2\right) = \frac{1}{\sqrt{2\pi\sigma^2}} \exp\left[-\frac{1}{2\sigma^2}\left(y - \mu\right)^2\right]$$

where y is any real number. Assuming that both μ and σ^2 are unknown parameters, $v = \mu$ and $\phi = \sigma^2$. Rewriting $f(y \mid \mu, \sigma^2)$ in exponential family form, we have

$$f\left(y \mid \mu, \sigma^2\right) = \exp\left[-\log\left(\sqrt{2\pi\sigma^2}\right) - \frac{1}{2\sigma^2}y^2 + \frac{1}{\sigma^2}y\mu - \frac{1}{2\sigma^2}\mu^2\right]$$

and we identify $t(y) = y$, $\eta(\mu) = \mu$, $a(\sigma^2) = \sigma^2$, $A(\mu) = \mu^2/2$, and

$$h(y, \sigma^2) = -\log\left(\sqrt{2\pi\sigma^2}\right) - y^2\Big/\left(2\sigma^2\right). \blacksquare$$

Table 2–1 contains a list of probability distributions belonging to the exponential family that are commonly used in generalized linear models. In addition to the exponential family of distributions, several other probability distributions are available for generalized linear modeling. These include the negative binomial, the non-central t, and the multinomial distributions (Table 2–2). If δ is known, the negative binomial belongs to the one parameter exponential family. The multinomial distribution generalizes the binomial distribution to more than two mutually exclusive and exhaustive categories. The categories can be either nominal (unordered) or ordinal (ordered or ranked).

TABLE 2.1. Examples of probability distributions that belong to the exponential family. All distributions, except for the log-normal distribution, have been parameterized such that $\mu = E(Y)$ is the mean of the random variable Y. For the log-normal distribution, the distribution of $Z = \log(Y)$ is normally distributed with mean $\mu_z = E[\log(Y)]$ and $\phi = \mathrm{var}[\log(Y)]$.

Distribution	$f(y \mid \mu)$	$\theta = \eta(\mu)$	Variance	ϕ
Normal (μ, ϕ) $-\infty < y < \infty$	$\dfrac{1}{\sqrt{2\pi\phi}}\exp\left[\dfrac{-(y-\mu)^2}{2\phi}\right]$	μ	ϕ	$\phi > 0$
Inverse normal (μ, ϕ) $-\infty < y < \infty$	$\left(\dfrac{1}{2\pi\phi y^3}\right)^{1/2}\exp\left[\dfrac{-(y-\mu)^2}{2y\phi\mu^2}\right]$	$1/\mu^2$	$\phi\mu^3$	$\phi > 0$
Log-normal (μ, ϕ) $-\infty < \log(y) < \infty$	$f\left[\log(y)\mid\mu\right] = \dfrac{1}{\sqrt{2\pi\phi}}\exp\left\{\dfrac{-\left[\log(y)-\mu\right]^2}{2\phi}\right\}$	μ	ϕ	$\phi > 0$
Gamma (μ, ϕ)† $y \geq 0$	$\dfrac{y^{\phi-1}}{\Gamma(\phi)}\left(\dfrac{\phi}{\mu}\right)^{\phi}\exp\left(\dfrac{-\phi y}{\mu}\right)$	$1/\mu$	$\phi\mu^2$	$\phi > 0$
Exponential (μ) $y \geq 0$	$\dfrac{1}{\mu}\exp\left(\dfrac{-y}{\mu}\right)$	$1/\mu$	μ^2	$\phi \equiv 1$
Beta (μ, ϕ)† $0 \leq y \leq 1$	$\dfrac{\Gamma(\phi)}{\Gamma(\mu\phi)\Gamma\left[(1-\mu)\phi\right]}y^{\mu\phi-1}\left(1-y\right)^{(1-\mu)\phi-1}$	$\log\left(\dfrac{\mu}{1-\mu}\right)$	$\dfrac{\mu(1-\mu)}{(1+\phi)}$	$\phi > 0$
Binomial (n, π) $y = 0, \ldots, n$ where $\pi = \mu/n$	$\dbinom{n}{y}\left(\dfrac{\mu}{n}\right)^{y}\left(1-\dfrac{\mu}{n}\right)^{n-y}$	$\log\left(\dfrac{\mu}{n-\mu}\right)$	$\mu\left(1-\dfrac{\mu}{n}\right)$	$\phi \equiv 1$
Geometric (μ, ϕ) $y = 0, 1, 2, \ldots$	$\left(\dfrac{\mu}{1+\mu}\right)^{y}\left(\dfrac{1}{1+\mu}\right)$	$\log(\mu)$	$\mu + \mu^2$	$\phi \equiv 1$
Poisson (μ)‡ $y = 0, 1, 2, \ldots$	$\dfrac{\mu^y e^{-\mu}}{y!}$	$\log(\mu)$	μ	$\phi \equiv 1$

† The gamma function $\Gamma(x)$ equals $(x - 1)!$ when x is an integer but otherwise equals $\int_0^\infty t^{x-1}e^{-t}dt$.

‡ In the case of an over-dispersed Poisson distribution, the variance of Y is $\phi\mu$ where $\phi > 0$ and often $\phi > 1$.

2.3 DESCRIPTIONS OF THE DISTRIBUTIONS

In this section, each of the non-normal distributions commonly used in generalized linear models is described, and examples of possible applications are given.

BETA

A random variable distributed according to the beta distribution is continuous, taking on values within the range 0 to 1. Its mean is μ, and its variance, $\mu(1 - \mu)/(1 + \phi)$, depends on the mean (Table 2–1). The beta distribution is useful for modeling proportions that are observed on a continuous scale in the interval (0, 1). The distribution is very flexible and, depending on the values of the parameters μ and ϕ,

TABLE 2-2. Additional probability distributions used in generalized linear models which do not belong to the one parameter exponential family of distributions. These distributions have been parameterized so that $\mu = E(Y)$ is the mean of the random variable Y.

Distribution	$f(y \mid \mu)$	$\theta = \eta(\mu)$	Variance	ϕ
Non-central t (v, μ, ϕ)† $-\infty < y < \infty,$ $v > 2$	$\dfrac{\Gamma\!\left(\frac{v+1}{2}\right)}{\Gamma\!\left(\frac{v}{2}\right)\phi\!\left(\frac{v-2}{v}\right)\sqrt{\pi v}}\left\{1 + v^{-1}\left[\dfrac{y-\mu}{\phi\!\left(\frac{v-2}{v}\right)}\right]^2\right\}^{-\left(\frac{v+1}{2}\right)}$	μ	$\phi^2\!\left(\dfrac{v-2}{v}\right)^2$	$\phi > 0$
Multinomial $(n,$ $P_1, P_2, \dots, P_k)$ $y_i = 0, 1, 2, \dots n,$ $i = 1, 2, \dots, k,$ $\sum_{i=1}^{k} y_i = n,$ where $p_i = \mu_i/n,$ $i = 1, 2, \dots, k$	$\begin{pmatrix} n \\ y_1, y_2, \dots, y_k \end{pmatrix}\prod_{i=1}^{k}\left(\dfrac{\mu_i}{n}\right)^{y_i}$	$\eta(\mu_i) = \log\!\left(\dfrac{\mu_i}{\mu_k}\right)$ $i = 1, 2, \dots, k-1$	$\mathrm{var}(y_i) = \mu_i\!\left(\dfrac{n - \mu_i}{n}\right)$ $i = 1, 2, \dots, k$	$\phi \equiv 1$
Negative binomial $(\mu,$ $\delta)$†‡ $y = 0, 1, 2, \dots,$ $\delta > 0$	$\dfrac{\Gamma\!\left(y+\delta^{-1}\right)}{\Gamma\!\left(\delta^{-1}\right)\Gamma\!\left(y+1\right)}\left(\dfrac{\mu}{\mu+\delta^{-1}}\right)^{y}\left(\dfrac{\delta^{-1}}{\mu+\delta^{-1}}\right)^{\delta^{-1}}$	$\log(\mu)$	$\mu + \dfrac{\mu^2}{\delta}$	—

† The gamma function $\Gamma(x)$ equals $(x-1)!$ when x is an integer but otherwise equals $\int_0^{\infty} t^{x-1}e^{-t}dt.$
‡ δ plays the role of the scale parameter but is not identically equal to $\phi.$

FIG. 2-1. Examples of the probability density function of a random variable having a beta distribution with parameters μ and ϕ.

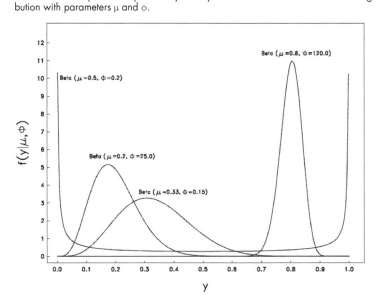

can take on shapes ranging from a unimodal, symmetric, or skewed distribution to a distribution with practically all of the density near the extreme values (Fig. 2–1).

Examples of the use of the beta distribution include modeling the proportion of the area in a quadrat covered in a noxious weed and modeling organic carbon as a proportion of the total carbon in a sample.

POISSON

A Poisson random variable is discrete, taking on non-negative integer values with both mean and variance μ (Table 2–1). It is a common distribution for counts per experimental unit, for example, the number of seeds produced per parent plant or the number of economically important insects per square meter of field. The distribution often arises in spatial settings when a field or other region is divided into equal sized plots and the number of events per unit area is measured. If the process generating the events distributes those events at random over the study region with negligible probability of multiple events occurring at the same location, then the number of events per plot is said to be Poisson distributed.

In many applications, the criterion of random distribution of events may not hold. For example, if weed seeds are dispersed by wind, their distribution may not be random in space. In cases of non-random spatial distribution, a possible alternative is to augment the variance of the Poisson distribution with a multiplicative parameter. The resulting "distribution" has mean μ and variance $\phi\mu$, where $\phi > 0$ and $\phi \neq 1$ but no longer satisfies the definition of a Poisson distribution. The word "distribution" appears in quotes because it is not a probability distribution but rather a quasi-likelihood (Section 2.5). It allows for events to be distributed somewhat evenly (under-dispersed, $\phi < 1$) over the study region or clustered spatially (over-dispersed, $\phi > 1$). When over-dispersion is pronounced, a preferred alternative to the scale parameter augmented Poisson quasi-likelihood is the negative binomial distribution that explicitly includes a scale parameter.

BINOMIAL

A random variable distributed according to the binomial distribution is discrete, taking on integer values between 0 and n, where n is a positive integer. Its mean is μ and its variance is $\mu[1 - (\mu/n)]$ (Table 2–1). It is the classic distribution for the number of successes in n independent trials with only two possible outcomes, usually labeled as success or failure. The parameter n is known and chosen before the experiment. In experiments with $n = 1$ the random variable is said to have a Bernoulli or binary distribution.

Examples of the use of the binomial distribution include modeling the number of field plots (out of n plots) in which a weed species was found and modeling the number of soil samples (out of n samples) in which total phosphorus concentration exceeded some prespecified level. It is not uncommon for the objectives in binomial applications to be phrased in terms of the probability or proportion of successes (e.g., the probability of a plot containing the weed species).

In some applications where the binomial distribution is used, one or more of the underlying assumptions are not satisfied. For example, there may be spatial correlation among field plots in which the presence or absence of a weed species was being recorded. In these cases, over-dispersion issues similar to those for the Poisson may arise.

NEGATIVE BINOMIAL

A negative binomial random variable is discrete, taking on non-negative integer values with mean μ and variance $\mu + \mu^2/\delta$, where δ ($\delta > 0$) plays the role of the scale parameter (Table 2–2). The negative binomial distribution is similar to the Poisson distribution in that it is a distribution for count data, but it explicitly incorporates a variance that is larger than its mean. As a result, it is more flexible and can accommodate more distributional shapes than the Poisson distribution.

Like the Poisson, the negative binomial is commonly used for counts in spatial settings especially when the events tend to cluster in space, since such clustering leads to high variability between plots. For example, counts of insects in randomly selected square-meter plots in a field will be highly variable if the insect outbreaks tend to be localized within the field.

The geometric distribution is a special case of the negative binomial where $\delta = 1$ (Table 2–1). In addition to modeling counts, the geometric distribution can be used to model the number of Bernoulli trials that must be conducted before a trial results in a success.

GAMMA

A random variable distributed according to a gamma distribution is continuous and non-negative with mean μ and variance $\phi\mu^2$ (Table 2–1). The gamma distribution is flexible and can accommodate many distributional shapes depending on the values of μ and ϕ. It is commonly used for non-negative and skewed response variables having constant coefficient of variation and when the usual alternative, a log-normal distribution, is ill-fitting.

The gamma distribution is often used to model time to occurrence of an event. For example, the time between rainfalls > 2.5 cm (>1 inch) per hour during a growing season or the time between planting and first appearance of a disease in a crop might be modeled as a gamma distributed random variable. In addition to time to event applications, the gamma distribution has been used to model total monthly rainfall and the steady-state abundance of laboratory flour beetle populations.

The exponential distribution is a special case of the gamma distribution where $\phi = 1$ (Table 2–1). The exponential distribution can be used to model the time interval between events when the number of events has a Poisson distribution.

LOG-NORMAL

A log-normal distributed random variable Y is a continuous, non-negative random variable for which the transformed variable $Z = \log(Y)$ is normally distributed with mean μ_Z and variance ϕ (Table 2–1). The untransformed variable Y has mean $\mu_Y = \exp(\mu_Z + \phi/2)$ and variance $\mathrm{var}(Y) = \exp(-\phi)\exp(\mu_Z + \phi/2)^2$. It is a common distribution for random variables Y which are continuous, non-negative, and skewed to the right but their transformed values $Z = \log(Y)$ appear to be normally distributed.

In addition, since the mean and variance of Y depend on the mean of $\log(Y)$, the variance of the untransformed variable Y increases with an increase in the mean.

The log-normal distribution can provide more realistic representations than the normal distribution for characteristics such as height, weight, and density, especially in situations where the restriction to positive values tends to create skewness in the data. It has been used to model the distribution of particle sizes in naturally occurring aggregates (e.g., sand particle sizes in soil), the average number of parasites per host, the germination of seed from certain plant species that are stimulated by red light or inhibited by far red light, and the hydraulic conductivity of soil samples over an arid region.

INVERSE NORMAL

An inverse normal random variable (also known as an inverse Gaussian) is continuous and non-negative with mean μ and variance $\phi\mu^3$. Like the gamma distribution, the inverse normal distribution is commonly used to model time to an event but with a variance larger than a gamma distributed random variable with the same mean.

NON-CENTRAL t

A non-central t distributed random variable is continuous over all real numbers with mean μ and variance $\phi^2 [(v - 2)/v]^2$, where v is a known constant, $v > 2$ (Table 2–1). The non-central t distribution is very similar in shape to the normal distribution, except that it has heavier tails than the normal distribution. The degree to which the tails are heavier than the normal distribution depends on the parameter v, commonly known as the degrees of freedom. When $\mu = 0$, the distribution is referred to as a central t or simply a t distribution.

The t distribution would be used as an alternative for the normal distribution when the data are believed to have a symmetric, unimodal shape but with a larger probability of extreme observations (heavier tails) than would be expected for a normal distribution. As a result of having heavier tails, data from a t distribution often appear to have more outliers than would be expected if the data had come from a normal distribution.

MULTINOMIAL

The multinomial distribution is a generalization of the binomial distribution where the outcome of each of n independent trials is classified into one of $k > 2$ mutually exclusive and exhaustive categories (Table 2–2). These categories may be nominal or ordinal. The response is a vector of random variables $[Y_1, Y_2, ..., Y_k]'$, where Y_i is the number of observations falling in the ith category and the Y_i sum to the number of trials n. The mean and variance of each of the Y_i are the same as for a binomially distributed random variable with parameters n and π, where the π_i sum to one and the covariance between Y_i and Y_j is given by $-n\pi_i\pi_j$.

The multinomial has been used to model soil classes that are on a nominal scale. It can also be used to model visual ratings such as disease severity or herbicide injury in a crop on a scale of one to nine. A multinomial distribution might also be used when n soil samples are graded with respect to the degree of infestation of nematodes into one of five categories ranging from none to severe.

2.4 LIKELIHOOD BASED APPROACH TO ESTIMATION

There are several approaches to estimating the unknown parameters of an assumed probability distribution. Although the method of least squares has been the most commonly used method for linear models where the response is normally distributed, the method has proven to be problematic for other distributions. An alternative approach to estimation that has been widely used is based on the likelihood concept.

Suppose that Y is a random variable having a probability distribution $f(y \mid \theta)$ that depends on an unknown parameter(s) θ. Let $Y_1, Y_2, \ldots, Y_n$ be a random sample from the distribution of Y. Because the Y values are independent, their joint distribution is given by the product of their individual distributions; that is,

$$f\left(y_1, y_2, \ldots, y_n \mid \theta\right) = f\left(y_1 \mid \theta\right) f\left(y_2 \mid \theta\right) \cdots f\left(y_n \mid \theta\right) = \prod_{i=1}^{n} f\left(y_i \mid \theta\right)$$

For a discrete random variable, the joint distribution is the probability of observing the sample $y_1, y_2, \ldots, y_n$ for a given value of θ. When thought of as a function of θ given the observed sample, the joint distribution is called the likelihood function and is usually denoted by $L(\theta \mid y_1, y_2, \ldots, y_n)$. From this viewpoint, an intuitively reasonable estimator of θ would be the value of θ that gives the maximum probability of having generated the observed sample compared to all other possible values of θ. This estimated value of θ is called the maximum likelihood estimate (MLE).

Assuming the functional form for the distribution of Y is known, finding maximum likelihood estimators is an optimization problem. Differential calculus techniques provide a general approach to the solution. In some cases, an analytical solution is possible; in others, iterative numerical algorithms must be employed. Since a non-negative function and its natural logarithm are maximized at the same values of the independent variable, it is often more convenient algebraically to find the maximum of the natural logarithm of the likelihood function.

EXAMPLE 2.3

Suppose that Y has a binomial distribution with parameters m and π. For a random sample of n observations from this distribution, the likelihood function is given by

$$L\left(\pi \mid y_1, y_2, \ldots, y_n\right) = \prod_{i=1}^{n} \binom{m}{y_i} \pi^{y_i} \left(1 - \pi\right)^{m - y_i}$$

The natural logarithm of the likelihood is

$$\log L\left(\pi \mid y_1, y_2, \ldots, y_n\right) = \sum_{i=1}^{n} \log\binom{m}{y_i} + \left(\sum_{i=1}^{n} y_i\right)\log(\pi) + \left(mn - \sum_{i=1}^{n} y_i\right)\log\left(1-\pi\right)$$

Differentiating $\log L(\pi \mid y_1, y_2, \ldots, y_n)$ with respect to π and setting the derivative equal to zero leads to

$$\left(\sum_{i=1}^{n} y_i\right)\left(\frac{1}{\pi}\right) - \left(mn - \sum_{i=1}^{n} y_i\right)\left(\frac{1}{1-\pi}\right) = 0$$

Solving for π yields the estimator

$$p = \frac{1}{mn}\sum_{i=1}^{n} y_i$$

Since the second derivative is negative, p maximizes the log-likelihood function. Hence, the sample proportion based on the entire sample is the maximum likelihood estimator of π. ∎

When Y is a continuous random variable, there are technical difficulties with the intuitive idea of maximizing a probability because, strictly speaking, the joint distribution (or probability density function) is no longer a probability. Despite this difference, the likelihood function can still be thought of as a measure of how "likely" a value of θ is to have produced the observed Y values.

EXAMPLE 2.4

Suppose that Y has a normal distribution with unknown mean μ and variance σ^2 so that $\theta' = [\mu, \sigma^2]$ is the vector containing both unknown parameters. For a random sample of size n, the likelihood function is given by

$$L\left(\theta \mid y_1, y_2, \ldots, y_n\right) = \prod_{i=1}^{n} \frac{1}{\sqrt{2\pi\sigma^2}}\exp\left[-\frac{1}{2\sigma^2}\left(y_i - \mu\right)^2\right]$$

and the log-likelihood is

$$\log L\left(\theta \mid y_1, y_2, \ldots, y_n\right) = -n\log\left(\sqrt{2\pi}\right) - n\log\left(\sigma^2\right) - \frac{1}{2\sigma^2}\sum_{i=1}^{n}\left(y_i - \mu\right)^2$$

Taking partial derivatives with respect to μ and σ^2, setting them equal to zero, and solving the resulting equations yields the estimators

$$\hat{\mu} = \bar{y} \quad \text{and} \quad \hat{\sigma}^2 = \frac{1}{n}\sum_{i=1}^{n}\left(y_i - \bar{y}\right)^2$$

Using the second partial derivatives, one can verify that these are the maximum likelihood estimators of μ and σ^2. Note that $\hat{\sigma}^2$ is not the usual estimator found in introductory textbooks where $1/n$ is replaced by $1/(n-1)$. We will return to this issue in Example 2.7 and in a more general context in Section 2.5. ∎

EXAMPLE 2.5

Suppose that Y has a gamma distribution with mean μ and scale parameter ϕ, so that $\theta' = [\mu, \phi]$. For a random sample of size n, the likelihood function is given by

$$L\left(\theta \mid y_1, y_2, \ldots, y_n\right) = \prod_{i=1}^{n} \frac{1}{\Gamma(\phi)} \left(\frac{\phi}{\mu}\right)^{\phi} y_i^{\phi-1} \exp\left(\frac{-\phi y_i}{\mu}\right)$$

and the log-likelihood is

$$\log L\left(\theta \mid y_1, y_2, \ldots, y_n\right) = -n \log \Gamma(\phi) + n\phi \log\left(\frac{\phi}{\mu}\right) + (\phi-1) \sum_{i=1}^{n} \log(y_i) - \frac{\phi}{\mu} \sum_{i=1}^{n} y_i$$

Because of the presence of the gamma function $\Gamma(\phi)$ in the distribution, no simple closed form solution for the maximum likelihood estimator of ϕ exists. Iterative numerical methods must be used to obtain it. ∎

Maximum likelihood estimators have the property that if $\hat{\theta}$ is an MLE of θ and $h(\theta)$ is a one-to-one function (i.e., $h(\theta_1) = h(\theta_2)$ if and only if $\theta_1 = \theta_2$), then the maximum likelihood estimator of $h(\theta)$ is $h(\hat{\theta})$. That is, the maximum likelihood estimator of a function of θ can be obtained by substituting $\hat{\theta}$ into the function. This result simplifies the estimation for parameters of interest derived from the basic parameters that define the distribution of Y.

EXAMPLE 2.6

In Example 2.3 the sample proportion p was shown to be the maximum likelihood estimator of π. Hence, the maximum likelihood estimator of the logit $\eta(\pi) = \log\left(\frac{\pi}{1-\pi}\right)$ is given by

$$\hat{\eta}(p) = \log\left(\frac{p}{1-p}\right) \quad ∎$$

In addition to being intuitively appealing, maximum likelihood estimators have many desirable theoretical properties. Under mild conditions, the method of maximum likelihood usually yields estimators that are consistent, asymptotically unbiased and efficient, and asymptotically normally distributed. For models with normally distributed data, likelihood based procedures can be shown to be equivalent to the more familiar least squares and analysis of variance based methods. For generalized and mixed models, likelihood based inference depends on asymptotic properties whose small sample behavior (like those typically found in much agricultural research) varies depending on the design and model being fit. As with any set of statistical procedures, there is no one-size-fits-all approach for maximum likelihood. More detailed discussions of these properties can be found in Pawitan (2001) and Casella and Berger (2002). When well-known estimation or inference issues that users should be aware of arise in examples in subsequent chapters, they will be noted and discussed in that context.

EXAMPLE 2.7

In Example 2.4, the maximum likelihood estimator of the variance of the normal distribution, σ^2, was shown to be

$$\hat{\sigma}^2 = \frac{1}{n}\sum_{i=1}^{n}\left(y_i - \bar{y}\right)^2$$

Recall that an estimator is unbiased if its mean (or expected value) is the parameter being estimated; that is, on average, the estimator gives the true value of the parameter. For $\hat{\sigma}^2$ the expected value is

$$E\left[\hat{\sigma}^2\right] = \left(\frac{n-1}{n}\right)\sigma^2 = \left(1 - \frac{1}{n}\right)\sigma^2$$

That is, the maximum likelihood estimator is a biased estimator of σ^2 with a bias of $-1/n$. For small sample sizes, the bias can be substantial. For example, for $n = 10$, the bias is 10% of the true value of σ^2. The negative bias indicates that the variance is underestimated, and hence, standard errors that use the estimator are too small. This leads to confidence intervals that tend to be too short, t and F statistics that tend to be too large, and, in general, results that appear to be more significant than they really are.

Note that the usual sample variance estimator taught in introductory statistical methods courses, namely,

$$S^2 = \frac{1}{n-1}\sum_{i=1}^{n}\left(y_i - \bar{y}\right)^2 = \left(\frac{n}{n-1}\right)\hat{\sigma}^2$$

has the expected value $E[S^2] = \sigma^2$; it is an unbiased estimator of σ^2. A common explanation given for the use of the denominator $n - 1$ instead of n is that one needs to account for having to estimate the unknown mean. ∎

2.5 VARIATIONS ON MAXIMUM LIKELIHOOD ESTIMATION

The concept of accounting for estimation of the mean when estimating the variance leads to a modification of maximum likelihood called residual maximum likelihood (REML). Some authors use the term restricted maximum likelihood as well. In Example 2.7, define the residuals $Z_i = Y_i - \bar{Y}$. The Z_i's have mean zero and variance proportional to σ^2. Hence, they can be used to estimate σ^2 independently of the estimate of μ. Applying maximum likelihood techniques to the Z_i's yields the REML estimator S^2 of σ^2; that is, the usual sample variance is a REML estimator.

In the context of linear mixed models, residual maximum likelihood uses linear combinations of the data that do not involve the fixed effects to estimate the random effect parameters. As a result, the variance component estimates associated with the random effects are independent of the fixed effects while at the same time taking into account their estimates. Details concerning the implementation of residual maximum likelihood can be found in Littell et al. (2006), Schabenberger and Pearce (2002), and McCulloch et al. (2008). For linear mixed models with normally distributed data, REML estimates are used almost exclusively because of the severe bias associated with maximum likelihood estimates for sample sizes typical of much agricultural research. For mixed models with non-normal data, REML is technically undefined because the existence of the residual likelihood requires independent mean and residuals, a condition only satisfied under normality. However, REML-like computing algorithms are used for variance-covariance estimation in non-normal mixed models when linearization (e.g., pseudo-likelihood) methods are used. Section 2.7 contains additional discussion of this issue.

For certain generalized linear models, the mean–variance relationship required for adequately modeling the data does not correspond to the mean–variance relationship of any member of the exponential family. Common examples include over-dispersion and repeated measures. Wedderburn (1974) developed the concept of quasi-likelihood as an extension of generalized linear model maximum likelihood to situations in which a model for the mean and the variance as a function of the mean can be specified. In addition, the observations must be independent. Quasi-likelihood is defined as a function whose derivative with respect to the mean equals the difference between the observation and its mean divided by its variance. As such the quasi-likelihood function has properties similar to those of a log-likelihood function. Wedderburn showed that the quasi-likelihood and the log-likelihood were identical if and only if the distribution of Y belonged to the exponential family. In general, quasi-likelihood functions are maximized using the same techniques used for maximum likelihood estimation. Details concerning the implementation of quasi-likelihood can be found in McCullagh and Nelder (1989) and McCulloch et al. (2008).

2.6 LIKELIHOOD BASED APPROACH TO HYPOTHESIS TESTING

Recall that we have a random sample $Y_1, Y_2, ..., Y_n$ from a random variable Y having a probability distribution $f(y \mid \theta)$ that depends on an unknown parameter(s) θ. When testing hypotheses concerning θ, the null hypothesis H_0 places restrictions on the possible values of θ. The most common type of alternative hypothesis H_1 in linear models allows θ its full range of possible values.

The likelihood function $L(\theta \mid y_1, y_2, ..., y_n)$ can be maximized under the restrictions in H_0 as well as in general. Letting $L(\hat{\theta}_0)$ and $L(\hat{\theta}_1)$ represent the maximum values of the likelihood under H_0 and H_1, respectively, the likelihood ratio

$$\lambda = L\left(\hat{\theta}_0\right)\Big/ L\left(\hat{\theta}_1\right)$$

can be used as a test statistic. Intuitively, if $L(\hat{\theta}_1)$ is large compared to $L(\hat{\theta}_0)$, then the value of θ that most likely produced the observed sample would not satisfy the restriction placed on θ by H_0 and, hence, would lead to rejection of H_0. The test procedure based on the ratio of the maximum values of the likelihood under each hypothesis is called a likelihood ratio test.

EXAMPLE 2.8

Suppose that Y has a normal distribution with unknown mean μ and unknown variance σ^2 so that $\theta' = [\mu, \sigma^2]$. Consider a test of the hypotheses

$H_0: \mu = \mu_0$ and $\sigma^2 > 0$ versus $H_1: \mu \neq \mu_0$ and $\sigma^2 > 0$

where μ_0 is a specified value. In the more familiar version of these hypotheses, only the mean appears since neither hypothesis places any restrictions on the variance. The reader may recognize this as a one sample t test problem. Here we consider the likelihood ratio test.

Under H_0, the mean is μ_0 so that the only parameter to be estimated is the variance σ^2. The maximum likelihood estimator of σ^2 given that the mean is μ_0 can be shown to be

$$\hat{\sigma}_0^2 = \frac{1}{n}\sum_{i=1}^{n}\left(y_i - \mu_0\right)^2$$

Under H_1, from Example 2.4 the maximum likelihood estimators are

$$\hat{\mu} = \bar{y} \text{ and } \hat{\sigma}_1^2 = \frac{1}{n}\sum_{i=1}^{n}\left(y_i - \bar{y}\right)^2$$

Substituting these estimators into the appropriate likelihoods, after some algebra the likelihood ratio reduces to

$$\lambda = \left[\frac{\sum_i\left(y_i - \mu_0\right)^2}{\sum_i\left(y_i - \bar{y}\right)^2}\right]^{n/2}$$

It can be shown that

$$\frac{\sum_i(y_i-\mu_0)^2}{\sum_i(y_i-\bar{y})^2} = \frac{\sum_i(y_i-\bar{y})^2 + n(\bar{y}-\mu_0)^2}{\sum_i(y_i-\bar{y})^2} = 1 + \frac{n(\bar{y}-\mu_0)^2}{\sum_i(y_i-\bar{y})^2} = 1 + \frac{n(\bar{y}-\mu_0)^2}{(n-1)s^2}$$

Note that the second term in the last expression is, up to a factor of $n - 1$, the square of the t statistic. Hence, the likelihood ratio test is equivalent to the usual one sample t test for testing the mean of a normal distribution. ∎

In Example 2.8 an exact distribution of the likelihood ratio statistic was readily determined. This is the case for all analysis of variance based tests for normally distributed data. When the exact distribution of the statistic is unknown or intractable for finite sample sizes, likelihood ratio tests are usually performed using $-2\log(\lambda)$ as the test statistic, where log is the natural logarithm. For generalized linear models, we use the result that the asymptotic distribution of $-2\log(\lambda)$ is chi-squared with v degrees of freedom, where v is the difference between the number of unconstrained parameters in the null and alternative hypotheses. Practically speaking, $-2\log(\lambda)$ having an asymptotic chi-squared distribution means that, for sufficiently large sample sizes, approximate critical values for $-2\log(\lambda)$ can be obtained from the chi-squared table. The accuracy of the approximation and the necessary sample size are problem dependent.

For one parameter problems, $(\hat{\theta} - \theta) / \sqrt{\text{var}_\infty(\hat{\theta})}$ is asymptotically normally distributed with mean zero and variance one, where $\hat{\theta}$ is the maximum likelihood estimator of θ and $\text{var}_\infty(\hat{\theta})$ is the asymptotic variance of $\hat{\theta}$. For normally distributed data, the asymptotic variance is often referred to as the "known variance." Because the square of a standard normal random variable is a chi-square, it follows that

$$W = \frac{\left(\hat{\theta} - \theta\right)^2}{\text{var}_\infty\left(\hat{\theta}\right)}$$

asymptotically has a chi-squared distribution with one degree of freedom. W is known as the Wald statistic and provides an alternative test procedure to the likelihood ratio test. More generally, for a vector of parameters θ, the Wald statistic is given by

$$W = \left(\hat{\theta} - \theta\right)' \left[\text{cov}_\infty\left(\hat{\theta}\right)\right]^{-1} \left(\hat{\theta} - \theta\right)$$

where $\text{cov}_\infty(\hat{\theta})$ is the asymptotic covariance matrix of $\hat{\theta}$. W has the same asymptotic chi-squared distribution as the likelihood ratio test.

EXAMPLE 2.9

Consider the one factor normal theory analysis of variance problem with K treatments and, for simplicity, n observations per treatment. The mean of the ith treatment can be expressed as $\mu_i = \mu + \tau_i$, subject to the restriction $\tau_1 + \ldots + \tau_K = 0$. The parameter μ is interpreted as the overall mean and the treatment effect τ_i as the deviation of the ith treatment mean from the overall mean. The initial hypothesis of equal treatment means is equivalent to

$H_0: \tau_1 = \ldots = \tau_K = 0$ versus $H_1:$ not all τ_i are zero.

The likelihood ratio statistic for testing H_0 is given by

$$\lambda = \left(\frac{SSE}{SSE + SSTrt} \right)^{Kn/2}$$

where SSTrt is the usual treatment sum of squares and SSE is the error sum of squares. We can rewrite λ as

$$\lambda = \left(\frac{1}{1 + \dfrac{SSTrt}{SSE}} \right)^{Kn/2} = \left[\frac{1}{1 + \dfrac{(K-1)}{K(n-1)} F} \right]^{Kn/2}$$

where F = MSTrt/MSE has an F distribution with $K - 1$ and $K(n - 1)$ degrees of freedom. Hence, the usual F-test in the analysis of variance is equivalent to the likelihood ratio test.

Because the maximum likelihood estimator of τ_i is the difference between the ith sample mean and the grand mean, it can be shown that the Wald statistic is given by

$$W = \frac{SSTrt}{\sigma^2}$$

where σ^2 is the common variance. Replacing σ^2 by its estimator MSE yields a test statistic for H_0; that is,

$$W = \frac{SSTrt}{MSE}$$

Note that W divided by the degrees of freedom associated with its numerator is the F statistic. This Wald statistic–F statistic relationship for the one factor problem will recur throughout generalized linear mixed models. ■

2.7 COMPUTATIONAL ISSUES

Parameter estimation and computation of test statistics increase in complexity as the models become more elaborate. From a computational viewpoint, linear models can be divided into four groups.

- Linear models (normally distributed response with only fixed effects): For parameter estimation, closed-form solutions to the likelihood equations exist and are equivalent to least squares. Exact formulas can be written for test statistics.
- Generalized linear models (non-normally distributed response with only fixed effects): The exact form of the likelihood can be written explicitly, as can the exact form of the estimating equations (derivatives of the likelihood). Solving the estimating equations to obtain parameter

estimates usually requires an iterative procedure. Likelihood ratio or Wald statistics can be computed for statistical inference.

- Linear mixed models (normally distributed response with both fixed and random effects): The exact form of the likelihood can be written explicitly as can the exact form of the estimating equations. There are two sets of estimating equations, one for estimating the model effects, commonly referred to as the mixed model equations and another for estimating the variance and covariance components. Solving the mixed model equations yields maximum likelihood estimates. These can be shown to be equivalent to generalized least squares estimates. The estimating equations for the variance and covariance are based on the residual likelihood; solving them yields REML estimates. Iteration is required to solve both sets of equations. Inferential statistics are typically approximate F or approximate t statistics. These can be motivated as Wald or likelihood ratio statistics, since they are equivalent for linear mixed models.

- Generalized linear mixed models (non-normally distributed response with both fixed and random effects): The likelihood is the product of the likelihood for the data given the random effects and the likelihood for the random effects, with the random effects then integrated out. Except for normally distributed data, the resulting marginal likelihood is intractable, and as a result, the exact form of the estimating equations cannot be written explicitly. Numerical methods such as those described below must be used. In theory, likelihood ratio statistics can be obtained. In practice, they are computationally prohibitive. Inference typically uses Wald statistics or approximate F statistics based on the Wald statistic.

Numerical techniques for finding MLEs and standard errors can be divided into two groups, linearization techniques and integral approximations. As the name implies, linearization uses a linear approximation to the log-likelihood, e.g., using a Taylor series approximation. This gives rise to a pseudo-variate that is then treated as the response variable of a linear mixed model for computational purposes. The mixed model estimating equations with suitable adjustments for the pseudo-variable and the associated estimating equations for variance and covariance components are solved. As with the linear mixed and generalized linear models, the solution process is iterative. Variations of linearization include pseudo-likelihood (Wolfinger and O'Connell, 1993) and penalized quasi-likelihood (Breslow and Clayton, 1993). The estimating equations for linear, linear mixed, and generalized linear models described above are all special cases of pseudo-likelihood.

The second group of techniques is based on integral approximations to the log-likelihood. This group includes the Laplace and Gauss–Hermite quadrature methods, Monte Carlo integration, and Markov chain Monte Carlo. The choice of a particular numerical method is problem dependent and will be discussed in the context of the various numerical examples in Chapter 5.

The most serious practical issue for iterative estimation procedures is convergence. Convergence is rarely a problem for generalized linear models and linear mixed models containing only variance components or at most simple covariance structures. However, as model complexity increases, the chance of encountering a convergence issue increases. The science and art of resolving convergence issues is an essential part of working with generalized and mixed models. Some convergence problems can be corrected easily by using different starting values or by increasing the number of iterations allowed before failure to converge is declared. In other cases, using a different algorithm may lead to convergence. Non-convergence may also result from ill-conditioned data; that is, data with very small or very large values or data ranging over several orders of magnitude. In these cases, a change of scale may eliminate the problem. Non-convergence also can result when there are fewer observations than parameters in the model being fit. This is especially possible for models having a large number of covariance parameters. Such problems require fitting a simpler model. In generalized linear mixed models non-convergence may be due to a "flat" likelihood function near the optimum. In extreme cases, it may be necessary to relax the convergence criterion to obtain a solution, although this should be considered a last resort.

2.8 FIXED, RANDOM, AND MIXED MODELS

Factors included in a statistical model of an experiment are classified as either fixed or random effects. Fixed factors or fixed effects are those in which the factor levels or treatments represent all of the levels about which inference is to be made. Fixed effects levels are deliberately chosen and are the same levels that would be used if the experiment were to be repeated. This definition applies to quantitative factors as well as qualitative effects; that is, in regression and analysis of covariance, the ranges of the observed values of the independent variables or covariates define the entire region to which inferences will apply. In contrast, random factors or random effects are those for which the factor levels in the experiment are considered to be samples from a larger population of possible factor levels. Ideally random effects levels are randomly sampled from the population of levels, and the same levels would not necessarily be included if the experiment were to be repeated. As a consequence of these definitions, fixed effects determine a model for the mean of the response variable and random effects determine a model for the variance.

Since the levels of a random factor are a sample (ideally random) from some population of possible factor levels and that population has an associated probability distribution, the random effects will also have a probability distribution. In general, it is assumed that the distribution of the random factor has a mean of zero and some unknown variance. For the mixed models discussed in this book, we further assume that random effects have normal distributions. In contrast, the factor levels of a fixed effect are a set of unknown constants.

In a given model an effect must be defined as either fixed or random. It cannot be both. However, there are certain types of effects that defy a one-size-fits-all categorization. Whether an effect is classified as fixed or random depends on how the factor levels were selected and the objectives of the experiment. For example, in a field experiment conducted at several locations, if the locations represent different crop growing regions and it is of interest to determine which of several cultivars are best suited for each region, then location would treated as a fixed effect. Inference would focus on differences among the location means. Moreover, inference is restricted to only those locations included in the experiment and cannot be extended to other locations. On the other hand, if the experiment included multiple locations to broaden the range of environmental conditions (e.g., weather, soil) under which the mean yields of the cultivars were being compared, then locations would be a random effect and inference would focus on the variability among locations. Treating locations as a random effect allows us to broaden inference to encompass the entire population of locations, not just those locations used in the experiment.

Usually fixed effects focus on the mean response and random effects focus on the variance. However, in the random location example, it may still be of interest to predict the yield of a particular cultivar at a particular location. In mixed models, this can be accomplished using best linear unbiased prediction (BLUP) that incorporates the random effects into the estimation of the mean for a particular location.

The model for a particular experiment is called a fixed effects model if all of the factors are fixed. A random effects model is one containing only random factors except for an intercept which is an unknown constant. If the model contains at least one fixed and at least one random effect, it is called a mixed model. In the early analysis of variance literature, fixed and random effects models were often referred to as model I and model II, respectively (Eisenhart, 1947).

Under the usual set of statistical assumptions for fixed effects models, the observed responses are assumed to be independent. However, this is not the case for random and mixed models. In these types of models, the random effects impose a correlation structure on the observations. For example, in a randomized complete block design with a fixed treatment factor and a random blocking effect, observations taken within the same block are correlated. Hence, analysis of data from experiments based on mixed models must take the correlation structure into account.

2.9 THE DESIGN-ANALYSIS OF VARIANCE-GENERALIZED LINEAR MIXED MODEL CONNECTION

The analysis of variance is arguably the fundamental tool for analyzing agronomic research data. Properly understood, analysis of variance (ANOVA) can be a valuable aid for understanding how to set up and work with generalized linear mixed models. Improperly understood, it can be a severe impediment to meaningful understanding of generalized linear mixed models and their role in agronomic research. Understanding the interconnections among ANOVA, design, and modeling is crucial for working effectively with generalized linear mixed models.

The analysis of variance was introduced by R.A. Fisher (Fisher and Mackenzie, 1923) in an article entitled "Studies in Crop Variation II: The Manorial Response of Different Potato Varieties." Once analysis of variance appeared, statistical scientists began attempts to place it in the framework of linear statistical models with varying degrees of success. Unfortunately, some of their lesser efforts still plague modern statistical practice in the experimental sciences. Speed (2010) described the uneasy relationship between statistical modeling and ANOVA, citing Fisher's remarks in the discussion accompanying Yates (1935). In those comments, Fisher (1935) described two aspects of the design of an experiment—topographical factors and treatment effects. Fisher used the word "topographical" because he was referring specifically to field experiments, but the term can be understood more broadly as the design structure that gives rise to all sources of variation in the observed data other than the treatments. Modern terminology refers to Fisher's topographical factors as the experiment design or design structure. Regardless of the terminology, the concept is important because, above all, a statistical model is a description, in mathematical and probabilistic terms, of the design and treatment factors and associated random variation giving rise to the observed data. Effective statistical modeling begins by asking "What would Fisher do?" and understanding his approach. Consider the following example as an illustration.

EXAMPLE 2.10

Suppose a field experiment is to be conducted to evaluate seed from two varieties of a certain crop using a randomized complete block design with 10 blocks. Data on several variables for each variety were taken according to the experiment's protocol. A schematic diagram of the experiment is shown in Table 2–3. In practice, we would randomize the order of the varieties within each block and follow any additional requirements of the design protocol. The schematic diagram serves mainly to show the essential design and treatment structures of the experiment.

The design structure for this experiment consists of the blocks and the two plots, one per variety, shown in Table 2–4. The sources of variation associated with the design structure are the variation among blocks and the variation between plots within each block (Table 2–5). The treatment structure is a single fixed factor consisting of the set of the two varieties. The associated source of variation is the variety effect with one degree of freedom. Integrating the design and

TABLE 2-3. Diagram of the seed evaluation experiment in Example 2.10.

Block	Variety	
1	A	B
2	A	B
3	A	B
4	A	B
5	A	B
6	A	B
7	A	B
8	A	B
9	A	B
10	A	B

TABLE 2-4. Diagram of the design structure for the seed evaluation experiment in Example 2.10.

Block	Plot	
1	–	–
2	–	–
3	–	–
4	–	–
5	–	–
6	–	–
7	–	–
8	–	–
9	–	–
10	–	–

TABLE 2-5. Sources of variation and degrees of freedom (df) for the design structure in Example 2.10.

Source of variation	df
Blocks	9
Plots within blocks	10
Total	19

TABLE 2-6. ANOVA table containing sources of variation and degrees of freedom (df) for the integrated design and treatment structures in Example 2.10.

Source of variation	df
Blocks	9
Varieties	1
Plots within blocks given varieties	9
Total	19

treatment structures yields the ANOVA table shown in Table 2–6. Note that the one degree of freedom for varieties is taken from the degrees of freedom for plots within blocks (the experimental unit to which varieties were randomly assigned), leaving nine degrees of freedom for plots after accounting for varieties. It is important to understand that when Fisher conceived ANOVA, the state of the art in statistical computing was little more than pencil and paper. Given this limitation, the practical way to assess the statistical significance of variety effects was to compare variation attributable to varieties as measured by MS(Variety) to naturally occurring variation associated with plots within blocks as measured by MS(WithinBlocks), more commonly referred to as MS(Error) or MS(Residual).

Up to this point, the analysis can be performed without reference to a statistical model. Proceeding further requires a statistical model. One well-known model assumes independent, normally distributed observations on each plot. The end result is an F-test using the ratio MS(Variety)/MS(WithinBlocks). What if one or both of these model assumptions is not true? For example, what if the response variable is binomial? Suppose in each plot we observe 100 plants of each variety and ask how many plants out of the 100 have a certain characteristic; for example, how many show evidence of damage from an insect pest or disease? This is where following Fisher's approach of identifying the experiment's processes becomes essential.

We begin by considering only the design structure processes.

- Design process 1: Variation among blocks. Let b_i denote the effect of the ith block, $i = 1, \ldots, 10$.

- Design process 2: Variation among plots within a block. Let y_{ij} denote the observation on the jth plot within the ith block, $i = 1, \ldots, 10, j = 1, 2$. Note that this is an observation on the plot, not an effect, because the plot is the unit on which the data are collected.

At this point, we specify any probability assumptions. If the blocks form a sample from a larger population that just as well could have consisted of any 10 blocks from this population (i.e., if blocks effects are random), then there is a probability distribution associated with the block effect. Linear mixed models and, in this book, generalized linear mixed models assume that the b_i are independent and normally distributed with mean zero and variance σ_B^2.

Observations on plots within blocks must be treated as random variables. Formally, in mixed model theory, each observation is conditional on the random effect level from which the observation arises. Denote this by $y_{ij} \mid b_i$ (the vertical bar is read as "given"). The conditional distribution has a conditional mean $\mu_{ij} \mid b_i$ and variance σ_W^2. If the conditional distribution of y_{ij} is normal, then we can express the conditional distribution of the observations as $y_{ij} \mid b_i \sim$ independent $N(\mu_{ij} \mid b_i, \sigma_W^2)$. Statistical modeling begins at this point.

Modeling consists of two steps:

- Decide on an equation that describes how the sources of variation affect μ_{ij}.
- Decide whether the mechanism described by this equation affects the mean μ_{ij} directly or indirectly; e.g., through the canonical parameter of the distribution.

For the normal distribution the natural or canonical parameter is the mean, and the decomposition is $\mu_{ij} \mid b_i = \mu + b_i + V_j$, where μ represents an intercept and V_j represents the effect of the jth variety (more generally, treatment). There is a long standing tradition of calling the parameter μ the overall mean. For generalized linear models, this becomes a dysfunctional habit that is important to break.

Now suppose that the observations are not normally distributed. For example, how do we form a model when the observation is the number of damaged plants out of the 100 plants observed per plot? In this case the distribution of the observations is $y_{ij} \mid b_i \sim$ independent Binomial($100, \pi_{ij}$), where π_{ij} denotes the probability that a plant in the ith block containing the jth variety shows evidence of damage. We still want to use $\beta_0 + b_i + V_j$ to characterize the block and variety effect on π_{ij}, where μ has been replaced by β_0 to reinforce the distinction between the intercept and the overall mean. There are several reasons not to use this decomposition to directly model π_{ij}, the most important being that if we do fit a model $\pi_{ij} = \beta_0 + b_i + V_j$ it is possible to obtain nonsensical estimates of the probability π_{ij} that are less than zero or greater than one. A better choice is to model the logit; i.e., $\log[\pi_{ij}/(1 - \pi_{ij})]$, which is the canonical parameter for the binomial distribution (Example 2.1). The resulting model is written as $\eta_{ij} = \log[\pi_{ij}/(1 - \pi_{ij})] = \beta_0 + b_i + V_j$. In generalized linear model terminology, η_{ij} is called the link function and $\beta_0 + b_i + V_j$ is called the linear predictor. ∎

Two important facts emerge from Example 2.10. First, the models for both the normal and binomial distributions use the same linear predictor, in this example, $\beta_0 + b_i + V_j$. Relating this to Table 2–6, the predictor is the additive list of the effects in the ANOVA, excluding the last line of the table. The last line refers to the unit of observation. How that line is incorporated into the model depends on the assumed probability distribution of the observations.

Second, for the normal distribution, when we estimate the effects in the linear predictor, we have an estimate of the mean $\mu_{ij} = \beta_0 + b_i + V_j$ but no information about the variance σ_W^2. We use the last line of the table to estimate this variance; i.e., $\hat{\sigma}_W^2$ = MS(WithinBlocks). This is where the tradition of referring to the last line of the ANOVA table as residual or error originates. It also means that we cannot include a block × treatment interaction in the linear predictor because it is confounded

TABLE 2-7. Types of linear models that may arise in the context of Example 2.10.

Distribution of observations	Example of the conditional distribution	Block effect	
		Fixed	Random (usually normally distributed)
Normal	$y_{ij} \mid b_i \sim N(\mu_{ij}, \sigma_W^2)$	Linear model	Linear mixed model
Non-normal	$y_{ij} \mid b_i \sim \text{Binomial}(100, \pi_{ij})$	Generalized linear model	Generalized linear mixed model

with the residual term required to estimate σ_W^2. In contrast, for the binomial distribution estimates of the model effects can be used to obtain an estimate of $\pi_{ij} = 1/\{1 + \exp[-(\beta_0 + b_i + V_j)]\}$. Since the variance for a binomial is $\pi_{ij}(1 - \pi_{ij})$, estimating π_{ij} allows us to estimate the variance as well. There is no separate σ_W^2, and hence, no separate estimate is required. This fundamentally changes the role of the last line of the ANOVA table. It is not residual and must not be thought of that way. Therefore, it is possible, and desirable in certain circumstances, to include a block × treatment interaction in the generalized linear model for the binomial. This point is not unique to the binomial and will resurface in several examples in Chapters 3 and 5.

Depending on the assumptions made by the researcher, we can distinguish among the four major types of linear models discussed in this book. These are shown in Table 2–7. An example of a normally distributed response would be the average seed weight of the 100 seeds. The proportion of seeds that germinated under a specified set of conditions may follow a binomial distribution, representing an example of a non-normally distributed response.

Example 2.10 illustrates the essential components of a linear model. They are:

- The conditional distribution of the response variable, Y, given the random effects embedded in the design process,

- The distribution of the random effects, often assumed to be normally distributed with mean zero and possibly with some general covariance structure,

- A link function applied to the conditional mean of the response variable,

- A linear predictor as a function of the design and treatment effects that is fit to the link function.

Working through the analysis of variance thought process of identifying design and treatment structures from first principles as Fisher envisioned them provides a coherent approach for constructing generalized linear mixed models. Example 2.10 illustrates that the four required components of the model arise naturally when working through this process.

Finally, a word of caution is in order. Textbooks dealing only with normally distributed observations typically give model equations containing an error term. For example, the model for Example 2.10 would be given as $y_{ij} = \mu + b_i + V_j + e_{ij}$. Writing a model in this way is only valid if the conditional distribution of the observations is normal and the link function is the identity function. Otherwise, as will be seen in subsequent chapters, the equations do not make sense. On the other hand, specifying a model using the essential elements above is valid for any linear model.

2.10 CONDITIONAL VERSUS MARGINAL MODELS

A distinction that arises in mixed models that does not occur in fixed effects models concerns conditional and marginal modeling. Models given by the four essential components listed in the previous section specify the conditional model—the name is derived from the fact that the distribution of the observations is specified conditionally on the random effects. Marginal models are an alternative way of specifying mixed models. As the name implies, they are specified in terms of the marginal distribution of the observations. The linear predictor of a marginal model contains only the fixed effects. The random effects are not modeled explicitly but their impact on variation is embedded in the covariance structure of the model.

For normally distributed data (linear mixed models), the distinction is more technical than consequential. Marginal models are useful for the analysis of repeated measures and as a way of accounting for negative variance component estimates. Chapter 4 contains examples illustrating conditional and marginal linear mixed models. For non-normally distributed data (generalized linear mixed models), the conditional versus marginal distinction is far more consequential because marginal models for non-normal data actually target different parameters than those we understand as we work through the model construction process described in the previous section.

Marginal models are usually called GEE-type models. The term GEE came from generalized estimating equation theory (Zeger and Liang, 1986; Liang and Zeger, 1986). Technically the term generalized linear mixed models (GLMM) refers only to conditional models. Chapter 5 begins with an illustration of the difference between conditional GLMMs and marginal GEE-type models. Other examples in Chapter 5 provide additional perspective.

2.11 SOFTWARE

Many statistical software packages can be used to analyze data from designed experiments. Only two of these, SAS (SAS Institute, Cary, NC) and R, will be described here. In our opinion, they represent the most widely used software packages in the applied statistics and agricultural sciences research communities in the United States.

The impetus for the creation of SAS came from a project in the 1970s sponsored by the Southern Region Agricultural Experiment Station directors to create a computer program to analyze data from designed experiments (Littell, 2011). Before 1990, the GLM procedure was the primary SAS tool for analyzing linear models with normally distributed responses. GLM was initially written for fixed effects models with the *random* statement added later to allow for random and mixed models. However, the fixed effect architecture of GLM severely limits the kinds of mixed models that can be accommodated and even with these models, its mixed model adjustments are limited. In the early 1990s, the MIXED procedure was introduced to make the full range of modeling options available for normally distributed response variables. It made the first row of Table 1–1 (normally distributed data) available to researchers and provided major improvements in both the ease and accuracy of the analysis of split plot, incomplete block, repeated measures, and spatial data for normally distributed responses.

The GENMOD procedure was introduced shortly after MIXED. This procedure made the full range of generalized linear model options available for data with fixed continuous or categorical explanatory variables as well as certain types of repeated measures and split plot analyses. However, GENMOD and MIXED together still left large portions of Table 1–1 inaccessible to researchers.

The recently introduced GLIMMIX procedure implements generalized linear mixed models. For a large range of distributions, it can address every explanatory model-response variable type combination in Table 1–1 under a common framework and a common syntax. This syntax is virtually identical to the syntax used by previous SAS procedures including GLM, MIXED, and GENMOD and thus is already familiar to most users.

R is a programming language for statistical computing and graphics that was designed in the early 1990s by R. Ihaka and R. Gentleman. Current developments and contributions are managed by the R Development Core Team. R is available for free as part of the GNU project. An initial installation includes a core set of packages, each containing functions to carry out specific tasks. In addition, a large number of user contributed packages on a wide variety of topics are available through the Comprehensive R Archive Network (CRAN). User contributed packages can address very specific analyses of interest to the contributor. For linear mixed models and generalized linear mixed models there are both fairly general as well as more narrowly focused packages. This collection of packages only loosely overlaps what is available in SAS. In cases where a SAS procedure may not be able to fit a model, there may be a suitable package available in R.

Linear mixed models with normally distributed responses and correlated error structures can be fit using the function *lme* in the R package *nlme*. This package also contains a function *nlme* that fits nonlinear mixed models. Several other packages are useful in special cases.

There are a number of packages in R that can fit various generalized linear models. The function *glm* in the package *stats* will fit linear models with binomial, gamma, normal, inverse normal, and Poisson responses. The function *glm.nb* in the package *MASS* allows for negative binomial distributed responses. The package *biglm* can be used for generalized linear regression with very large data sets. The functions *loglin* and *loglm* in the package *MASS* fit log-linear models with binomial, multinomial, and Poisson responses.

The package *MASS* contains the function *glmmPQL*, which fits a generalized linear mixed model with multivariate normal random effects and most of the response distributions found in SAS GLIMMIX. This function uses penalized quasi-likelihood. The function *lmer* in the package *lme4* will also fit generalized linear mixed models using either a Laplace or Gauss–Hermite approximation to the log-likelihood and has easier to use syntax than *glmmPQL*. However, it does not produce F-tests and *p*-values. These must be calculated separately within R or elsewhere by the user. Alternatively, one can obtain confidence intervals on parameter estimates from their posterior distributions (created using Markov chain Monte Carlo methods) using the *mcmcsamp* function in the *lme4* package. The model estimation method used by *lmer* does not allow for modeling correlation at the individual observation level. Modeling correlation at the individual observation level requires a penalized quasi-likelihood approach. The function *glmm.admb* in the package *glmmADMB* fits generalized linear mixed models for binomial, Poisson, and negative binomial distributed responses with Laplace approximation as the default algorithm for likelihood estimation. The package *glmmML* can be used to fit binomial and Poisson responses with fixed covariates and random intercepts. The *repeated* package contains approximately 20 functions for fitting specialized generalized linear mixed models, many containing repeated measures.

In addition to the abovementioned packages and functions, there are a number of other R packages that fit various generalized linear mixed models using other numerical techniques. There are also packages within R that take a Bayesian approach to model fitting and estimation.

In this book, we use SAS exclusively and within SAS, almost all of the examples use PROC GLIMMIX. Readers are cautioned that the computational issues discussed in Section 2.7 still remain with this procedure as well as with any of the R packages. Blind acceptance of the default options of a procedure is not recommended and may lead to completely unreasonable results. Boykin et al. (2011) demonstrated the issues that arise when options are used that are not appropriate for the problem at hand. Examples in Chapters 4 and 5 contain specific examples of overrides of defaults essential to complete an appropriate analysis.

REFERENCES CITED

Boykin, D., M.J. Camp, L.A. Johnson, M. Kramer, D. Meek, D. Palmquist, B. Vinyard, and M. West. 2011. Generalized linear mixed model estimation using PROC GLIMMIX: Results from simulations when the data and model match, and when the model is misspecified. p. 137–170. *In* W. Song (ed.) Conference Proceedings on Applied Statistics in Agriculture. Department of Statistics, Kansas State University, Manhattan, KS.

Breslow, N.E., and D.G. Clayton. 1993. Approximate inference in generalized linear mixed models. J. Am. Stat. Assoc. 88:9–25. doi:10.2307/2290687

Casella, G., and R.L. Berger. 2002. Statistical inference. Duxbury Press, North Scituate, MA.

Eisenhart, C. 1947. The assumptions underlying the analysis of variance. Biometrics 3:1–21. doi:10.2307/3001534

Fisher, R.A. 1935. The design of experiments. Oliver and Boyd, Edinburgh, UK.

Fisher, R.A., and W.A. Mackenzie. 1923. Studies in crop variation II: The manurial response of different potato varieties. J. Agric. Sci. 13:311–320. doi:10.1017/S0021859600003592

Liang, K.-Y., and S.L. Zeger. 1986. Longitudinal data analysis using generalized linear models. Biometrika 73:13–22. doi:10.1093/biomet/73.1.13

Littell, R.C. 2011. The evolution of linear models in SAS®: A personal perspective. *In* Proceedings of SAS Global Forum 2011. Paper 325-2011. SAS Inst., Cary, NC.

Littell, R.C., G.A. Milliken, W.W. Stroup, R.D. Wolfinger, and O. Schabenberger. 2006. SAS for mixed models. 2nd ed. SAS Inst., Cary, NC.

McCullagh, P., and J.A. Nelder. 1989. Generalized linear models. 2nd ed. Chapman and Hall, New York.

McCulloch, C.E., S.R. Searle, and J.M. Neuhaus. 2008. Generalized, linear and mixed models. 2nd ed. John Wiley and Sons, New York.

Pawitan, Y. 2001. In all likelihood. Statistical modeling and inference using likelihood. Oxford Univ. Press, Oxford, UK.

Schabenberger, O., and F.J. Pearce. 2002. Contemporary statistical models for the plant and soil sciences. CRC Press, Boca Raton, FL.

Speed, T. 2010. Terence's stuff: An ANOVA thing. IMS Bull. 39:16.

Wedderburn, R.W.M. 1974. Quasi-likelihood functions, generalized linear models and the Gauss–Newton method. Biometrika 61:439–447.

Wolfinger, R., and M. O'Connell. 1993. Generalized linear mixed models: A pseudo-likelihood approach. J. Stat. Comput. Simul. 48:233–243. doi:10.1080/00949659308811554

Yates, F. 1935. Complex experiments. Suppl. J. R. Stat. Soc. 2:181–247. doi:10.2307/2983638

Zeger, S.L., and K.-Y. Liang. 1986. Longitudinal data analysis for discrete and continuous outcomes. Biometrics 42:121–130. doi:10.2307/2531248

GENERALIZED LINEAR MODELS

3.1 INTRODUCTION

Generalized linear models extend normal theory linear models to response variables whose distributions belong to the exponential family or can be characterized by a quasi-likelihood (Section 2.5). This class of models includes fixed effects analysis of variance as well as regression and analysis of covariance models that do not contain random effects. A generalized linear model consists of three components:

- a stochastic component that defines the probability distribution or quasi-likelihood of the response variable,

- a linear predictor that is a systematic component describing the linear model defined by the explanatory variables,

- a link function that relates the mean of the response variable to a linear combination of the explanatory variables.

More specifically, let Y be the response variable with a probability distribution from the exponential family with mean $E[Y] = \mu$ and let $x_1,..., x_p$ be a set of p explanatory variables. For the jth observation from a random sample of size n, the systematic component can be expressed as

$$\eta_j = g\left(\mu_j\right) = \beta_0 + \sum_{i=1}^{p} \beta_i x_{ij}, j = 1,...,n$$

where $g(\cdot)$ is the link function, μ_j is the mean of the jth observation, and x_{ij} is the observed value of the ith explanatory variable for the jth observation.

Writing the above system of equations in matrix form,

$$\eta = g(\mu) = X\beta$$

where $g(\mu)' = [g(\mu_1),..., g(\mu_n)]'$ is the $n \times 1$ vector of link functions, X is the $n \times (p + 1)$ design matrix, and β is the $(p + 1) \times 1$ vector of unknown parameters. It is important to emphasize that a generalized linear model involves relating the transformed mean of the response variable Y to the explanatory variables but does not involve

doi:10.2134/2012.generalized-linear-mixed-models.c3

Analysis of Generalized Linear Mixed Models in the Agricultural and Natural Resources Sciences
Edward E. Gbur, Walter W. Stroup, Kevin S. McCarter, Susan Durham, Linda J. Young, Mary Christman, Mark West, and Matthew Kramer

transforming the response variable itself. Hence, the data remain on the original scale of measurement (data scale), but the model for the mean as a linear function of the explanatory variables is on a different scale (link scale or model scale). This approach is not the same as transforming the data to a different scale from the original scale of measurement. For example, application of the log transformation for counts followed by a normal theory based analysis of variance is not the same as a generalized linear model assuming a Poisson distribution and log link.

The linear predictor component of a generalized linear model creates an intrinsically linear relationship between a function of the mean of the response and the explanatory variables. Valid link functions are monotone (i.e., either increasing or decreasing) and are differentiable. If $g(\cdot)$ has the same functional form as the canonical parameter $\theta = \eta(\cdot)$ of the distribution of Y, it is referred to as the canonical link. Table 3–1 lists commonly used link functions for the distributions described in Section 2.3.

TABLE 3-1. Commonly used link functions for the probability distributions described in Section 2.3.

Distribution	SAS default link function	Other available link functions
Normal	μ	—
Inverse normal	$1/\mu^2$	—
Lognormal	μ	—
Non-central t	μ	—
Gamma	$\log(\mu)$	$1/\mu$
Exponential	$\log(\mu)$	$1/\mu$
Beta	$\log[\mu/(1-\mu)]$	—
Binomial†	$\log[\mu/(n-\mu)]$	(1) $\log[-\log(1-\mu/n)]$ (2) $-\log[-\log(\mu/n)]$ (3) $\Phi^{-1}(\mu/n)$
Geometric	$\log(\mu)$	—
Poisson	$\log(\mu)$	—
Multinomial with ordinal categories†‡	$\log[\pi_i/(1-\pi_i)]$ for ith category	(1) $\log[-\log(1-\pi_i)]$ (2) $-\log[-\log(\pi_i)]$ (3) $\Phi^{-1}(\pi_i)$
Multinomial with nominal categories§	$\log(\mu_i/\mu_k)$ for ith category	—
Negative binomial	$\log(\mu)$	—

† $\Phi^{-1}(p) = x$ is the quantile function of the standard normal distribution where x is the $100p$th quantile. This link function is usually referred to as the probit link. The link function $\log[-\log(1-\mu/n)]$ is referred to as the complementary log-log link and the function $-\log[-\log(\mu/n)]$ as the log-log link.

‡ $\pi_i = \sum_{j=1}^{i}\left(\mu_j/n\right)$ is the cumulative probability for the first i ordered categories.

§ The kth category is considered as the base category against which the others are compared.

The link function is important because it allows hypothesis testing and related inference to be performed using existing statistical methods such as t-tests and F-tests on the link scale, which is linear in the β's. Since the response variable is not transformed, its assumed distribution provides the basis for the maximum likelihood estimation and inference. The monotonicity of the link function guarantees that the inverse link function, $g^{-1}(\cdot)$, exists and provides a vehicle for transforming results from the link scale back to the scale on which the response Y was observed (the data scale).

3.2 INFERENCE IN GENERALIZED LINEAR MODELS

Parameters in generalized linear models are estimated using maximum likelihood. Numerical algorithms that can be used to obtain the estimators and their estimated standard errors for those models in which closed form algebraic solutions do not exist for all parameters were described in Section 2.7. Hypothesis tests are constructed using Wald-type statistics based on the maximum likelihood estimators. Since both the maximum likelihood estimators and the Wald statistics are asymptotically normal, many software packages report approximate tests and confidence intervals based on the t and F distributions. In addition to Wald-type tests, likelihood ratio tests can be used. Regardless of the particular procedure, all inference is performed on the transformed or link scale of the mean rather than the original or data scale of the response variable.

EXAMPLE 3.1

Let Y be the response variable and let $x_1,..., x_p$ be a set of explanatory variables. Suppose that Y has a Poisson distribution with mean μ and that the canonical link

$$\eta = \log(\mu) = \beta_0 + \sum_{i=1}^{p} \beta_i x_i$$

is used to relate μ to the explanatory variables. Then the predicted values $\hat{\eta}$ from the model fitting will be on the log scale, and the reported standard errors are for those predicted values. In a regression model, each estimated coefficient would be interpreted in terms of a change in $\log(\mu)$ for a unit increase in the associated explanatory variable when all other explanatory variables are kept fixed. In an analysis of variance model it would be interpreted as the deviation of $\log(\mu)$ from the intercept for the level of the associated factor. ∎

Estimation of contrasts is also performed on the link scale. For example, in an analysis of variance model, the difference between two factor levels, i and j, is estimated as the difference between $g(\mu_i)$ and $g(\mu_j)$. Both the point estimate and the standard error of the difference are determined on the link scale. The endpoints of a confidence interval for the difference on the link scale can be constructed as the point estimate plus or minus a critical value times the estimated standard error. All hypothesis testing of contrasts is performed on the link scale.

To report predicted values on the original scale requires converting the estimates from the link scale to the data scale using the inverse of the link function, $\mu = g^{-1}(\eta)$. This process is referred to as inverse linking. In addition to inverse linking the estimates, their standard errors must be inverse linked as well. These are obtained using a technique known as the delta method for approximating the variance of a function of a random variable. A description of the technique and conditions under which it is accurate are given in Oehlert (1992) and Agresti (2002).

It is important to realize that the inverse linked estimators $\hat{\mu} = g^{-1}(\hat{\eta})$ of the means on the data scale are not necessarily equal to the sample means calculated from the original data. This has ramifications for presentation of the results from an analysis based on a generalized linear model. When discrepancies occur, it is often because the sample mean is not an appropriate measure of central tendency for the assumed distribution of the observations. Hence, the inverse linked estimated means and not the sample means of the original data are the appropriate results to be reported.

EXAMPLE 3.2

Let Y be the response variable and let $x_1, ..., x_p$ be a set of explanatory variables. Suppose that Y has a binomial distribution with mean $\mu = n\pi$. Then $\pi = \mu/n$, where π is the probability of observing a success, and the logit link can be expressed as

$$\eta = \log\left[\pi/(1-\pi)\right] = \beta_0 + \sum_{i=1}^{p} \beta_i x_i$$

Applying the inverse link,

$$\frac{\pi}{1-\pi} = e^{\eta}$$

Solving for π yields

$$\pi = \frac{e^{\eta}}{1+e^{\eta}}$$

The maximum likelihood estimator (MLE), p, of π is obtained by substituting the MLE of β into the above equation. Since p is the MLE, it is asymptotically normally distributed. The asymptotic mean is

$$E[p] = \frac{e^{\pi}}{1+e^{\pi}} = \frac{e^{\mu/n}}{1+e^{\mu/n}}$$

Using the delta method, the asymptotic variance is found to be

$$\mathrm{var}[p] = \frac{e^{\pi}}{n\pi(1-\pi)(1+e^{\pi})^2} = \frac{e^{\mu/n}}{\mu(1-\mu/n)(1+e^{\mu/n})^2} \quad \blacksquare$$

For confidence intervals for individual means on the original scale of the data, inverse linking of the endpoints of the confidence interval constructed on the link scale is recommended rather than inverse linking the mean and standard error and then constructing a symmetric confidence interval. The first approach yields a confidence interval having the same coverage probability as the confidence interval on the link scale but leads to intervals that are not symmetric about the mean for most link functions. An asymmetric interval is appropriate on the data scale when the probability distribution of the response variable is not symmetric.

The inverse link function can be applied to means on the link scale but in general not directly to pairwise differences or other contrasts. If it is of interest to report the estimated differences or linear combinations of means on the data scale, one cannot simply apply the inverse link to the difference estimated on the link scale because of the nonlinear form of the link functions used in generalized linear models. The appropriate method is to inverse link the means on the link scale and then take the difference of the resulting data scale estimates. Obtaining the estimated standard errors of the contrasts requires approximating the estimated standard errors of the means using the delta method and then combining them for the linear combination of interest in the same manner as would be used for linear models with normally distributed data.

EXAMPLE 3.3

Suppose we are interested in pairwise comparisons of means in a one factor analysis of variance model with three levels where the response variable Y has a Poisson distribution with mean μ_i for the ith factor level. For the canonical link function, the model becomes $\log(\mu_i) = \beta_0 + \beta_i$, where β_0 is the intercept on the link scale and β_i is the deviation of the ith level mean from the intercept, also on the link scale. The difference between two means on the link scale is $\log(\hat{\mu}_i) - \log(\hat{\mu}_j)$ which inverse links to

$$e^{\log(\hat{\mu}_i) - \log(\hat{\mu}_j)} = e^{\log(\hat{\mu}_i/\hat{\mu}_j)} = \hat{\mu}_i/\hat{\mu}_j$$

that is, the inverse link converts the link scale difference to the ratio of the estimated means and not to their difference.

The data scale estimator of the mean for each level is $\hat{\mu}_i = \exp\left(\hat{\beta}_0 + \hat{\beta}_i\right)$ and their difference is given by $\hat{\mu}_i - \hat{\mu}_j = \exp\left(\hat{\beta}_0 + \hat{\beta}_i\right) - \exp\left(\hat{\beta}_0 + \hat{\beta}_j\right)$. The standard error of this difference would be approximated by $\sqrt{\mathrm{v\hat{a}r}\left(\hat{\mu}_i\right) + \mathrm{v\hat{a}r}\left(\hat{\mu}_j\right)}$, where $\mathrm{v\hat{a}r}\left(\hat{\mu}_i\right)$ is approximated using the delta method.■

It is important to understand that each of the procedures described above is based on the approximate normality of the probability distribution of the estimators. While this should be true asymptotically (i.e., for sufficiently large sample sizes), it is not necessarily the case given the smaller sample sizes used in many agricultural experiments. Hence, the stated coverage probability of the confidence interval may or may not be accurate, and the results should be viewed with caution when the sample sizes are small. The sample sizes needed for reasonably accurate approximations have not been fully examined in the statistics literature. Sample size and related issues are briefly discussed in Section 3.4 and are the focus of Chapter 7.

The next example provides an illustration of the application of generalized linear models to an actual experiment. It also introduces the SAS (SAS Institute, Cary, NC) version 9.2 GLIMMIX procedure.

EXAMPLE 3.4

Improving nitrogen (N) fertilizer management includes consideration of the soil's ability to supply N to the plant. Soil testing methods such as the Illinois Soil Nitrogen Test (ISNT) and the direct steam distillation method (DSD) have been suggested as predictors of potentially mineralizable N. Bushong et al. (2008) conducted a study of N recovery by the two methods for a large number of soils of varying textures. In this example, we analyze some related unpublished data provided by R.J. Norman and T.L. Roberts (used with permission) for six soils, three from each of two texture classes (clay and silt loam). Only data from the 0- to ~15-cm (0- to 6-inch) depth are included. For each method, three samples of each soil were spiked with a fixed amount of ^{15}N-glucosamine and the proportion of glucosamine recovered was recorded. The proportion recovered for one sample exceeded one because of measurement error and was not included in the analysis, resulting in 35 observations.

Soil texture class (*texture*) and analysis method (*method*) are fixed effects. For the purpose of this example, we will treat the soil effects (*soil*) as fixed; i.e., as if we are only interested in inference for these six soils. If the soils were considered as a random sample from a large population of soils, then the soil effects would be random. Since each soil belongs to a specific texture class, soil is nested within texture class. Method and texture have a factorial treatment structure since all combinations of method and texture are present in the experiment. Method and soil have a factorial structure as well.

Since the response Y is the proportion recovered, it is restricted to values between zero and one. Before the development of generalized linear models, proportions measured on a continuous scale (i.e., not a binomial proportion) were often analyzed as if they were normally distributed or were arcsine-square root transformed and treated as normally distributed on the transformed scale. A disadvantage of analyzing proportions assuming a normal distribution is that estimated proportions based on the normal model can be negative or larger than one.

In this example, we assume that Y has a beta distribution that is by definition restricted to the interval from zero to one (Section 2.3) and will use the logit link $\eta = g(\mu) = \log[\mu/(1-\mu)]$. Based on the design of the experiment, the model for $E(Y) = \mu$ is given by

$$\eta_{ijk} = \log\left(\frac{\mu_{ijk}}{1 - \mu_{ijk}}\right) = \beta_0 + T_i + S_{j(i)} + M_k + TM_{ik} + SM_{jk(i)}$$

for $i = 1, 2$; $j = 1, 2, 3$; $k = 1, 2$,

where β_0 is the intercept, T_i is the ith texture effect, $S_{j(i)}$ is the jth soil effect within the ith texture, M_k is the kth method effect, and TM_{ik} and $SM_{jk(i)}$ are the texture × method and method × soil within texture interaction effects, respectively.

The model can be written in matrix form as

$$\eta = g(\mu) = X\beta$$

where β contains the 27 fixed effects parameters (intercept + 2 texture effects + 6 soil effects + 2 method effects + 4 texture × method interaction effects + 12 method × soil within texture interaction effects) and X is the 35 × 27 design matrix.

The GLIMMIX statements used to fit this model are given in Fig. 3–1. The proportion recovered is denoted by *prop*. Note that the distribution and link are specified as options on the *model* statement. The output in Fig. 3–2 contains basic information about the model and estimation procedure. The default variance function referred to in the *Model Information* section is var(Y) = $\mu(1 - \mu)/(1 + \phi)$ for the beta distribution (Table 2–1). The diagonal variance matrix indicates that the observations are uncorrelated. The covariance parameter referred to in the "Dimensions" section is the scale parameter ϕ. The 27 columns in X in that section match the parameter count obtained above.

The results of the F-tests for the fixed effects are presented in Fig. 3–3. Since these F-tests are based on maximum likelihood estimators and Wald statistics and not on least squares, the sums of squares and means squares found in conventional analysis of variance tables for normal theory linear models are not relevant here. Based on the *p*-values (*Pr > F* column), the texture class and methods main effects and the soils within texture class effects are significant at the 0.05 level.

FIG. 3–1. GLIMMIX statements to fit the beta model for Example 3.4.

```
proc glimmix data=soil plots=(residualpanel pearsonpanel studentpanel);
class texture soil method;
model prop = texture soil(texture) method texture*method method*soil(texture) /
    dist=beta link=logit;
lsmeans texture method / ilink adjust=tukey;
lsmeans soil(texture) / ilink slice=texture slicediff=texture
    plot=meanplot(sliceby=texture join cl) adjust=tukey;
output out=new pred(ilink)=predi stderr(ilink)=sepredi pred=pred stderr=sepred
    resid=resid student=student;
quit;
```

FIG. 3-2. GLIMMIX output containing basic model and fitting information for Example 3.4.

Model Information

Data Set	Work Soil
Response Variable	prop
Response Distribution	Beta
Link function	Logit
Variance Function	Default
Variance Matrix	Diagonal
Estimation Technique	Maximum Likelihood
Degrees of Freedom Method	Residual

Class Level Information

Class	Levels	Values
texture	2	Clay Siltloam
soil	6	ARShark Beaumont Forestdale Ganado LA-RP LA-VP
method	2	DSD ISNT

Number of Observations Read	36
Number of Observations Used	35

Dimensions

Covariance Parameters	1
Columns in X	27
Columns in Z	0
Subjects (Blocks in V)	1
Max Obs per Subject	35

FIG. 3-3. GLIMMIX output containing F-tests for the fixed effects for Example 3.4.

Type III Tests of Fixed Effects

Effect	Num DF	Den DF	F Value	Pr > F
texture	1	23	5.13	0.0332
soil(texture)	4	23	3.07	0.0366
method	1	23	115.98	<0.0001
texture*method	1	23	0.74	0.3996
soil*method(texture)	4	23	2.30	0.0890

The least squares means for the texture and method main effects produced by the first *lsmeans* statement in Fig. 3–1 are presented in Fig. 3–4 along with tests for the differences. The columns labeled *Estimate* and *Standard Error* contain the

FIG. 3-4. GLIMMIX output containing least squares means and comparisons for texture and methods for Example 3.4.

Texture Least Squares Means

Soil texture	Estimate	Standard Error	DF	t Value	Pr > |t|	Mean	Standard Error Mean
Clay	2.8924	0.05485	23	52.74	<0.0001	0.9475	0.002730
SiltLoam	2.7185	0.05421	23	50.15	<0.0001	0.9381	0.003147

Differences of texture Least Squares Means
Adjustment for Multiple Comparisons: Tukey–Kramer

Soil texture	Soil texture	Estimate	Standard Error	DF	t Value	Pr > |t|	Adj P
Clay	SiltLoam	0.1738	0.07672	23	2.27	0.0332	0.0332

Method Least Squares Means

Measurement method	Estimate	Standard Error	DF	t Value	Pr > |t|	Mean	Standard Error Mean
DSD	2.3916	0.04333	23	55.19	<0.0001	0.9162	0.003327
ISNT	3.2192	0.06386	23	50.41	<0.0001	0.9616	0.002361

Differences of Method Least Squares Means
Adjustment for Multiple Comparisons: Tukey–Kramer

Measurement method	Measurement method	Estimate	Standard Error	DF	t Value	Pr > |t|	Adj P
DSD	ISNT	−0.8276	0.07685	23	−10.77	<0.0001	<0.0001

estimated linear predictor and its standard error on the logit scale. The *t value* and *Pr > |t|* columns give the t-statistic and its *p*-value for a test of whether or not the mean on the logit scale is zero. The *ilink* option on the *lsmeans* statement adds the inverse linked estimates converted from the logit scale to the proportion (or data) scale and their associated standard errors to the output. The estimated standard errors on the data scale are based on the delta method. These statistics are found in the columns labeled *Mean* and *Standard Error Mean*.

As an illustration of the difference between inverse linking standard errors and calculating approximate standard errors using the delta method, for clay textured soils, $\hat{\eta}_{1\cdot\cdot}$ = logit $(\hat{\mu}_{1\cdot\cdot})$ = 2.8924 and $\hat{\mu}_{1\cdot\cdot}$ = exp(2.8924)/[1 + exp(2.8924)] = 0.9475. Inverse linking the estimated standard error of 0.0549 on the logit scale to the proportion scale would yield 0.5137, while the delta method produces an approximate estimated standard error of 0.0027 on the data scale.

In the *Difference of Least Squares Means* section, the *adjust* = *tukey* option on the *lsmeans* statement produces the *p*-value for the test of whether or not the difference is zero on the logit scale using the Tukey–Kramer procedure described in Kramer (1956). The *p*-value based on the Tukey–Kramer procedure is given in the column labeled *Adj P*. By default, the t-statistic and its *p*-value are reported regardless of the method requested by the *adjust* option. The t-test is equivalent to the least significant difference (LSD) method. When there are only two levels of the factor, the *p*-values for the two procedures are identical. For factors with more than two levels the Tukey–Kramer procedure *p*-values are larger than the corresponding *p*-values for the LSD since the Tukey–Kramer method controls the experiment-wise error rate.

The second *lsmeans* statement provides the analysis of the soil means within each texture. Since soils are nested within texture class, the *slice* = *texture* option requests a test for equality of soil means within each texture class separately. These are sometimes referred to as tests of simple effects (Winer et al., 1991). Figure 3–5 contains the results of these F-tests. For example, the *p*-value of 0.0462 for clay-textured soils indicates that not all three of the clay soils have the same effect on the logit scale at the 0.05 significance level.

The *slicediff* = *texture* option requests estimates of pairwise differences between least squares means of soils within their texture class. As a result, a clay soil would not be compared to a silt loam soil. The least squares means and their differences within texture classes are shown in Fig. 3–6. As before, the adjusted *p*-values are based on the Tukey–Kramer procedure. For clay soils, the adjusted *p*-values do not indicate any differences in the means, apparently a contradiction to the F-test result in Fig. 3–5. Based on the unadjusted *p*-values from LSD procedure, the ARShark soil differs significantly from the other two clay soils (p = 0.0251 and 0.0491), which are not significantly different (p = 0.7653).

In trying to sort out these results several things must be kept in mind. First, except for Scheffé's procedure, there is no theoretical link between the F-test and any of the multiple comparison procedures (Scheffé, 1999). In addition, Scheffé's procedure only guarantees that there will be at least one contrast that is significant when the F-test is significant. This contrast may not be a simple difference between a pair of means. In fact, the significant contrast may not be of any practical interest. Second, the chosen significance level for the Tukey–Kramer procedure represents the experiment-wise error rate, which forces the comparison-wise (or individual)

FIG. 3–5. GLIMMIX output containing F-tests for soil differences within each texture class for Example 3.4.

Tests of Effect Slices for soil(texture) Sliced By texture				
Soil texture	Num DF	Den DF	F Value	Pr > F
Clay	2	23	3.53	0.0462
SiltLoam	2	23	2.61	0.0953

FIG. 3-6. GLIMMIX output containing least squares means and comparisons for soils within each texture class for Example 3.4.

soil(texture) Least Squares Means

Soil texture	Soil series	Estimate	Standard Error	DF	t Value	Pr > |t|	Mean	Standard Error Mean
Clay	ARShark	2.6971	0.08670	23	31.11	<0.0001	0.9369	0.005129
Clay	Beaumont	3.0109	0.09851	23	30.57	<0.0001	0.9531	0.004406
Clay	LA-RP	2.9690	0.09821	23	30.23	<0.0001	0.9512	0.004563
SiltLoam	Forestdale	2.5892	0.08389	23	30.86	<0.0001	0.9302	0.005450
SiltLoam	Ganado	2.6727	0.09014	23	29.65	<0.0001	0.9354	0.005447
SiltLoam	LA-VP	2.8937	0.1055	23	27.43	<0.0001	0.9475	0.005243

Simple Effect Comparisons of soil(texture) Least Squares Means By Texture

Simple Effect Level	Soil series	Soil series	Estimate	Standard Error	DF	t Value	Pr > |t|	Adj P
texture Clay	ARShark	Beaumont	-0.3138	0.1310	23	-2.40	0.0251	0.0626
texture Clay	ARShark	LA-RP	-0.2718	0.1308	23	-2.08	0.0490	0.1165
texture Clay	Beaumont	LA-RP	0.04192	0.1388	23	0.30	0.7653	0.9511
texture SiltLoam	Forestdale	Ganado	-0.08349	0.1229	23	-0.68	0.5039	0.7779
texture SiltLoam	Forestdale	LA-VP	-0.3045	0.1346	23	-2.26	0.0334	0.0817
texture SiltLoam	Ganado	LA-VP	-0.2210	0.1385	23	-1.60	0.1242	0.2677

error rate to be much smaller. Hence, procedures such as Tukey–Kramer, which control the experiment-wise error rate, tend to find fewer significant differences, but fewer of these differences tend to be false differences. In contrast, the LSD procedure controls the comparison-wise error rate at the expense of a larger experiment-wise error rate. Hence, it tends to find more differences, but more of these differences tend to be false differences. Ultimately the choice of a procedure depends on which error rate is more important to control in the particular subject matter context. The books by Miller (1981) and Hochberg and Tamhane (1987) discuss the statistical issues involved in multiple comparisons. The articles by Carmer and Swanson (1973), Chew (1976), Baker (1980), and Day and Quinn (1989) provide some guidance on selection of a procedure.

The *meanplot* option on the second *lsmeans* statement creates a graph of the estimated soil means on the logit scale (Fig. 3–7). The vertical scale label may be somewhat confusing, but the range of numerical values clearly indicates the logit scale is being used. If the *ilink* option had been added within *meanplot*, the inverse linked means on the proportion (data) scale would have been graphed. The vertical axis label for the proportion scale would have read *inverse link proportion recovered*. The *cl* option adds confidence intervals about each mean. The *sliceby =*

FIG. 3-7. GLIMMIX output displaying a graph of the least squares means on the logit scale for soils within texture class for Example 3.4.

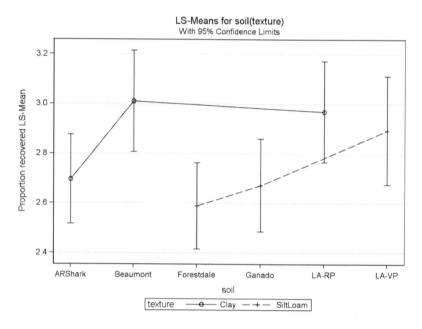

texture option in conjunction with the *join* option connects the means within each texture class by straight lines. While this type of plot is informative and is often used in publications, it should not be used alone to determine statistically significant differences among means. In general, nonoverlapping confidence intervals do not indicate a significant difference, nor do overlapping intervals indicate no significant difference. Care must be taken when making inferences based on the plot alone (Schenker and Gentleman, 2001; Ryan and Leadbetter, 2002).

3.3 DIAGNOSTICS AND MODEL FIT

As in linear models, it is important to check whether or not the data satisfy the model assumptions. Diagnostic statistics and model checking methods used for linear models provide a starting point for generalized linear models. The departure of individual observations from the model can be checked using many of the same methods that are used in linear models. In analysis of variance type generalized linear models the form of the linear predictor component is determined by the design of the experiment and is usually not subjected to model checking. For regression and the covariate portion of analysis of covariance models where the form of the linear predicator is being driven by a search for a model with good predictive ability, the form should be evaluated for appropriateness. In all cases, model checking should address the appropriateness of the stochastic component that defines the probability distribution of the response variable, the link function, and the variance function.

Informal model checking methods are typically graphical in nature, relying on visual assessment of pattern. Graphical analysis of residuals, which exists in many forms, is the centerpiece of informal methods.

One approach to formal model checking methods embeds the current model in a wider class of models that may include additional parameters in the linear predictor, other distributions for the response variable, different links, and/or different variance functions. The current model would be preferred to other models in the class if they do not appreciably improve model fit compared to the current model (McCullagh and Nelder, 1989). Another useful approach, especially in situations where predictive ability is important, embeds the current model in a class of models that have simpler forms of the systematic component. Another model in the class would be preferred to the current model if it does not provide an appreciably worse fit than the current model.

McCullagh and Nelder (1989, p. 414) note that "model checking remains as much art as science." Other authors have described it as a science based art form. Departures from a model may be due to any number of factors, such as an incorrect choice of the link function, the wrong scale for an explanatory variable, a missing interaction term, an outlier or influential observation, or a typographical error in the data. Multiple departures can interact in complex ways, often making an iterative approach necessary for careful model checking.

GOODNESS OF FIT

The Pearson chi-square statistic and the deviance statistic can be used to assess the overall fit of a generalized linear model. The Pearson chi-square statistic is given by

$$X^2 = \sum_{i=1}^{n} \frac{\left(y_i - \hat{\mu}_i \right)^2}{\text{var}\left(\hat{\mu}_i \right)}$$

The deviance statistic is twice the difference between the log-likelihood in which the means are replaced by the observed responses and the log-likelihood evaluated at the $\hat{\mu}_i$ (i.e., the difference between $-2 \log(L)$ for the full data and $-2 \log(L)$ for the fitted model). The deviance generalizes the sum of squared errors in normal theory linear models. For sufficiently large sample sizes, both statistics have approximate chi-square distributions with $n - d$ degrees of freedom, where n is the sample size and d is the number of estimated parameters. Large values of either statistic indicate lack of fit. The deviance statistic is often preferred because it provides a likelihood based inference and can be used for comparing nested models (Gill, 2001).

Goodness of fit among competing models also can be assessed using various information criteria statistics. In addition to comparing alternative models having the same probability distribution, these statistics are of particular use in determining the correct probability distribution for a generalized linear model, given the fixed effects in the model. Among the more popular criteria are:

- Akaike's Information Criterion (Akaike, 1974): AIC $= -2\,\mathrm{log}L\!\left(\hat{\theta}\mid y\right) + 2d$,
- AICC (Hurvich and Tsai, 1989): AICC $=$ AIC $+ 2d(d+1)/(n-d-1)$,
- Bayesian Information Criterion (Schwarz, 1978):
  BIC $= -2\,\mathrm{log}L\!\left(\hat{\theta}\mid y\right) + d\,\mathrm{log}\!\left(n\right)$,

where $\mathrm{log}L\!\left(\hat{\theta}\mid y\right)$ is the maximum value of the log likelihood (the likelihood evaluated at the maximum likelihood estimator $\hat{\theta}$) and d is the number of parameters in the model when the observations are independent. AICC is a bias-corrected small sample version of AIC, and Schwartz's criterion is similar to AIC but has a larger penalty for the number of parameters. Smaller values of a criterion indicate a better fitting model.

Although information criteria statistics can be used for comparing certain probability distributions for the response variable (Burnham and Anderson, 2002), they should be used with caution. For such comparisons to be valid, all aspects of the models except for the choice of distribution must be kept constant. Models to be compared must use the same observations, the same explanatory variables, and the same response variable. In addition, information criteria for models fit using pseudo-likelihood or quasi-likelihood that approximate the original model by a linear model should not be compared to models fit with techniques such as the Laplace and quadrature methods that are based on the log-likelihood.

For example, these information criteria can be used to compare the appropriateness of the Poisson and negative binomial as candidate distributions for the response when fit to the same set of data with the same set of explanatory variable terms. Similarly, they can be used to compare the gamma and normal distributions for the response. In contrast, when comparing the log-normal and gamma distributions, if the log-normal is fit by first transforming the response Y to a new response $\mathrm{log}(Y)$, which then is fit using a normal distribution (as is done by the GLIMMIX procedure), the information criteria cannot be compared since the response variables are not the same in both fits.

In regression type models where the form of the systematic component is not predetermined but is driven by the predictive ability of the model as the primary objective, information criteria can be used to compare the fit of a full model to a reduced model in which one or more of the explanatory variable terms have been removed. As above, the data and distributional assumptions must remain unchanged.

RESIDUALS

Residuals represent the difference between the data and the model. As such, they play a central role in model checking, particularly in informal graphical methods. For generalized linear models, residuals can be defined on both the data scale as $y - \hat{\mu} = y - g^{-1}\!\left(\hat{\eta}\right)$ and the link scale as $p - \hat{\eta}$, where p denotes a pseudo-data value that arises from the linearization algorithm used to obtain the maximum likelihood estimator (Section 2.7).

The difference $y - \hat{\mu} = y - g^{-1}\!\left(\hat{\eta}\right)$ is usually referred to as the raw residual or simply, residual. The Pearson residual for an observation is the signed square root of its contribution to the Pearson chi-square statistic and can be defined on both the link and data scales. The deviance residual for an observation is the signed square root of its contribution to the deviance statistic. Studentized residuals are defined on either scale by dividing the residual by the square root of its estimated variance. Studentized residuals have constant variance in contrast to Pearson and deviance residuals, which do not. Both Pearson and studentized residuals are approximately normally distributed on the link scale. The studentized deviance residual is preferred for model checking procedures because its distributional properties are closer to those of residuals from the normal theory linear model (Faraway, 2006; Gill, 2001; McCullagh and Nelder, 1989).

All types of residuals can be plotted against a variety of statistics and indices, but in some cases the resulting graph may not be very informative. For example, residual plots are not useful for distributions with a limited number of different observed response values such as binary responses, binomial responses when the number of trials n is small, and Poisson responses with small means. In these cases, residual plots will show curved lines of points that correspond to the observed response values.

CHECKING FOR ISOLATED DEPARTURES

Individual observations may be inconsistent with the model due to, for example:

- an error in recording the observation,
- an error in selecting certain sample units from the population,
- lack of homogeneity in the population for the explanatory variables under consideration,
- observation of a rare but possible expression of the phenomenon under study,
- unforeseen aspects of the phenomenon under study, resulting in an insufficiently specified model.

Observations with extreme values of the response and/or explanatory variables are unusual only relative to the model fit to them. If a discrepant observation is correct as recorded and the departure is deemed important, then a decision must be made about whether or not and how to modify the model to accommodate the departure. In general, final decisions should be subject-matter based.

Isolated departures can be detected using the leverage and/or Cook's distance for each observation. Particularly large values will be apparent in index plots of each statistic versus the observation number. In probability and quantile plots, observations that fit the model poorly will occur at plot extremes and typically fail to follow a trend established by other observations.

McCullagh and Nelder (1989) described a formal test of isolated departures in which an indicator variable is created, taking a value of zero for discrepant observations and a value of one otherwise. The effect of the discrepant observations on model fit can be assessed using either the change in deviance or a likelihood ratio test. This indicator variable approach is equivalent to comparing models fit with and without the discrepant observations. Cautious interpretation of significance is recommended because the discrepant observations are subjectively selected.

CHECKING FOR SYSTEMATIC DEPARTURES

Systematic departures can result from incorrect assumptions about the stochastic and/or systematic components of the model. The model also may fail to fit the data as a whole because of an incorrect choice of the link function. For regression type models where good predictive ability by the model is the objective, lack of fit could be due to missing explanatory variables or to incorrect functions of explanatory variables such as scale and interactions in the linear model defining the linear predictor of the generalized linear model. Systematic departures may also result from missing random effects and lack of independence of observations. These random elements usually should be addressed after violations of isolated departures and issues associated with the systematic component have been resolved.

As with linear models, plots of residuals against some function of the fitted values may uncover violations of model assumptions. Studentized deviance or Pearson residuals should be plotted against the linear predictor $\hat{\eta}$ on the link scale or against predicted means $\hat{\mu}$ on the data scale. Augmenting the plot with a smoothed curve can be helpful in some cases. The null pattern for this plot is a distribution of studentized residuals with mean zero and constant range for varying $\hat{\eta}$ or $\hat{\mu}$. Typical systematic departures are the appearance of curvature in the mean or systematic change in the range of residual values as the fitted value changes; i.e., nonconstant variance. Curvature may indicate an incorrect choice of link function, an incorrect choice of scale for one or more explanatory variable or omission of a function of an explanatory variable (e.g., a quadratic term). Nonconstant variance may suggest an incorrect choice of distribution or variance function.

For continuous scale explanatory variables in regression and analysis of covariance models, plots of studentized residuals on the link scale versus individual explanatory variables should have the same null pattern described above. Systematic departure may indicate an incorrect choice of scale for the explanatory variable which should have a linear relationship with the response on the link scale, an incorrect choice of link function, omission of a function of the explanatory variables or an incorrect scale for another explanatory variable that is closely related. For categorical scale predictors, side by side boxplots of studentized residuals for each level of the explanatory variable permit examination of the constant variance assumption.

For regression models, an added variable plot can be used to determine if an omitted potential explanatory variable should be incorporated in the linear predictor $\eta(\cdot)$ (McCullagh and Nelder, 1989; Draper and Smith, 1998). The potential explanatory variable is used as the response variable in a generalized linear model having the same form as the model for Y. The raw (non-studentized) residuals from the fitted model for Y are plotted against the raw residuals from the fitted model for the potential explanatory variable. The null pattern is no trend in the plot, indicating that a potential explanatory variable was appropriately omitted.

Uncertainty about the link function is most common for the gamma and binomial distributions. A plot of the response Y transformed to the link scale versus $\hat{\eta}$ can be used to examine the adequacy of the choice of the link function. The null pattern is a straight line. A simple formal test involves adding $\hat{\eta}^2$ as an additional explanatory variable in the model. The current link function is adequate if $\hat{\eta}^2$ fails to appreciably improve the fit as determined by the change in deviance or a likelihood ratio test. Alternatively, a family of link functions in which the current link is a member can be defined, a range of links can be fit, and goodness of fit statistics can be compared. McCullagh and Nelder (1989) and Faraway (2006) provide additional details.

EXAMPLE 3.5

This example is a continuation of Example 3.4 focusing on assessment of the adequacy of the beta model that was fit to the data.

Various goodness of fit statistics are displayed in Fig. 3–8. These are part of the default output of GLIMMIX. While smaller values of the negative of the log-likelihood and the information criteria indicate a better fit, their magnitude can be interpreted only relative to the fit of an alternative model. Hence, they are not of much value by themselves. The Pearson chi-square goodness of fit statistic was 36.06. Calculation of the p-value based on a chi-square distribution with 23 degrees of freedom gave 0.0407, which is marginally significant.

Unlike PROC MIXED in SAS, the *model* statement in GLIMMIX does not have an option that will produce a SAS dataset or printed list of predicted values, residuals and related diagnostics. Instead, the *output* statement or *ods* tables can be used to create a dataset containing these statistics. For each statistic to be included in the file, the appropriate keyword followed by an equal sign and a variable name is added to the *output* statement. The *output* statement in Fig. 3–1 creates the dataset *new* containing the original variables, the predicted values and their standard errors on both the logit and proportion scales, and the raw residuals and the studentized residuals

FIG. 3–8. GLIMMIX output containing various measures of goodness of fit for the model in Example 3.4.

Fit Statistics	
–2 Log Likelihood	–213.23
AIC (smaller is better)	–187.23
AICC (smaller is better)	–169.89
BIC (smaller is better)	–167.01
CAIC (smaller is better)	–154.01
HQIC (smaller is better)	–180.25
Pearson Chi-Square	36.06
Pearson Chi-Square / DF	1.57

FIG. 3-9. SAS statements to print predicted values and residuals for Example 3.5.

```
proc print data=new noobs label; by texture soil method;
   id texture soil method;
   var rep prop logit pred sepred resid student;
   format logit pred sepred resid student 7.4;
quit;
```

on the logit scale. Figure 3–9 contains the SAS statements for printing selected variables from the output file. Figure 3–10 contains the printed listing.

The *plots* option on the PROC GLIMMIX statement in Fig. 3–1 creates a set of plots for the raw residuals, the Pearson residuals, and the studentized residuals. By default, the plots are presented as a 2 × 2 panel of plots. The panel consists of a scatter plot of residuals versus the linear predictor, a histogram of the residuals with a normal density overlaid, a Q-Q or quantile plot, and a boxplot of the residuals. The *unpack* option (not shown) produces the plots individually instead of in panel form.

The studentized residual panel is shown in Fig. 3–11. The boxplot indicates the possibility of a slightly skewed distribution although the relatively small sample size (35 observations) used to construct the plot prevents a definitive conclusion. Both the boxplot and quantile plot indicate the possibility of one or more outliers. From Fig. 3–10, the first sample from the *LA-VP* soil series measured by the *DSD* method is the most outlying observation. The other observations that should be checked for correctness are the second *DSD* sample from Forestdale and the first and third *ISNT* observations from Ganado.

As a final note, the *plots* option on the PROC GLIMMIX statement can be used to produce the least squares means plots that were created by the *plots* option on the *lsmeans* statement. We recommend that these plots should be created from the *lsmeans* statement since it allows more control over the format of the plots. ■

3.4 GENERALIZED LINEAR MODELING VERSUS TRANSFORMATIONS

An often-asked question is why a generalized linear model is needed if one could instead do a transformation of the response variable so that the transformed data are approximately normally distributed. This was a common approach before the development of computer algorithms for estimating parameters of linear models with non-normally distributed response variables. In fact, there is a large body of literature on the use of transformations, going at least as far back as Fisher's (1921) transformation of the correlation coefficient r, $z = \text{arctanh}(r)$, to obtain approximate normality. These studies were largely motivated by the fact that non-normal distributions have variances that are functions of the means and so are by definition heterogeneous, but the transformed variables had stable variances. Further, many transformations were also shown to produce symmetric distribu-

FIG. 3-10. GLIMMIX output containing the list of predicted values and residuals for Example 3.5.

Soil texture	Soil series	Measurement method	Sample	Proportion recovered	Logit	Linear Predictor	Std. Error Linear Predictor	Residual	Studentized Residual
Clay	ARShark	DSD	1	0.9145	2.3699	2.3069	0.1009	0.0613	0.0176
			2	0.9032	2.2333	2.3069	0.1009	−0.0759	−0.0218
			3	0.9134	2.3559	2.3069	0.1009	0.0480	0.0138
Clay	ARShark	ISNT	1	0.9697	3.4658	3.0873	0.1409	0.3195	0.0654
			2	0.9456	2.8555	3.0873	0.1409	−0.2580	−0.0528
			3	0.9537	3.0252	3.0873	0.1409	−0.0639	−0.0131
Clay	Beaumont	DSD	1	0.9493	2.9298	2.6857	0.1182	0.2197	0.0538
			2	0.9236	2.4923	2.6857	0.1182	−0.2105	−0.0515
			3	0.9365	2.6911	2.6857	0.1182	0.0054	0.0013
Clay	Beaumont	ISNT	1	0.9577	3.1197	3.3362	0.1574	−0.2396	−0.0437
			2	0.9709	3.5075	3.3362	0.1574	0.1583	0.0289
			3	0.9704	3.4899	3.3362	0.1574	0.1432	0.0261
Clay	LA-RP	DSD	1	0.9175	2.4089	2.5418	0.1112	−0.1407	−0.0367
			2	0.9348	2.6629	2.5418	0.1112	0.1150	0.0300
			3	0.9310	2.6022	2.5418	0.1112	0.0588	0.0153
Clay	LA-RP	ISNT	1	0.9694	3.4557	3.3962	0.1617	0.0578	0.0103
			2	0.9689	3.4390	3.3962	0.1617	0.0419	0.0074
			3	0.9680	3.4095	3.3962	0.1617	0.0132	0.0023
SiltLoam	Forestdale	DSD	1	0.9079	2.2883	2.1201	0.0938	0.1574	0.0488
			2	0.8620	1.8320	2.1201	0.0938	−0.3224	−0.0999
			3	0.9065	2.2716	2.1201	0.0938	0.1427	0.0442
SiltLoam	Forestdale	ISNT	1	0.9493	2.9298	3.0582	0.1390	−0.1362	−0.0282
			2	0.9627	3.2507	3.0582	0.1390	0.1765	0.0366
			3	0.9559	3.0762	3.0582	0.1390	0.0178	0.0037
SiltLoam	Ganado	DSD	1	0.8902	2.0928	2.0332	0.0907	0.0582	0.0187
			2	0.8732	1.9296	2.0332	0.0907	−0.1079	−0.0346
			3	0.8915	2.1062	2.0332	0.0907	0.0709	0.0227
SiltLoam	Ganado	ISNT	1	0.9833	4.0755	3.3121	0.1557	0.5441	0.1004
			2	0.9576	3.1173	3.3121	0.1557	−0.2135	−0.0394
			3	0.9453	2.8496	3.3121	0.1557	−0.5761	−0.1063
SiltLoam	LA-VP	DSD	1	0.8978	2.1730	2.6620	0.1170	−0.6058	−0.1499
			2	0.9495	2.9340	2.6620	0.1170	0.2419	0.0598
			3	0.9495	2.9340	2.6620	0.1170	0.2419	0.0598
SiltLoam	LA-VP	ISNT	1	0.9634	3.2704	3.1254	0.1754	0.1358	0.0273
			2	1.0027	.	3.1254	0.1754	.	.
			3	0.9543	3.0389	3.1254	0.1754	−0.0900	−0.0181

FIG. 3-11. GLIMMIX output containing graphs of the studentized residuals for Example 3.5.

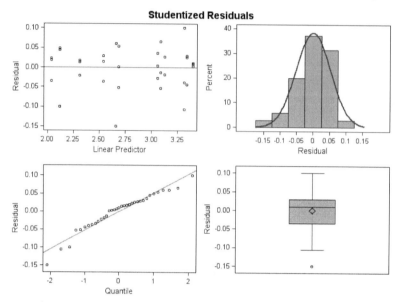

tions simultaneously that are better approximated by a normal distribution. These transformations have a long history in analysis of variance (e.g., Curtiss, 1943; Fisher, 1954). However, they can be problematic for regression settings in which the transformation also affects the functional relationship between the explanatory variables and the response variable.

Sometimes it is not recognized that the use of transformations changes the model under consideration. McArdle and Anderson (2004) discussed the example of the log transformation, which is often used for skewed distributions to obtain symmetry. In a symmetric distribution, the mean and median are the same so that a confidence interval for the mean of the log-transformed data is also a confidence interval for the median. Back-transforming the confidence interval endpoints yields a confidence interval for the median on the original scale because the monotonicity of the logarithm and exponential functions preserves the order of the data. However, the mean of the data on the log scale is not equal to the log of the mean of the original data; e.g., $[\log(Y_1) + \log(Y_2)]/2 \neq \log[(Y_1 + Y_2)/2]$. Hence, the back-transformed confidence interval does not provide information about the mean on the original scale.

More generally, for the comparison of means from skewed distributions, the back-transformation of a confidence interval for the difference in means on the log scale will produce a confidence interval for the ratio of the medians on the original scale. It does not give information on the difference between the means on the original scale. An exception arises when the original data are log-normally distributed and the groups have a common variance. Generalized linear models avoid these issues because the data are not transformed; instead, a function of the means is modeled as a linear combination of the explanatory variables.

In some cases, the use of a transformation has been shown to be more effective than using generalized linear models and Wald type statistics for inference. For example, in logistic regression, the Wald type test of a coefficient in the model uses $\chi^2 = \hat{\beta}^2 / \mathrm{var}\left(\hat{\beta}\right)$, which is asymptotically chi-square distributed. Menard (2002) stated that for large values of the estimated coefficient, the denominator is inflated, leading to small test statistic values and hence, less likely rejection of the null hypothesis. Conversely, the arcsine-square root transformation of binomial proportions for stabilizing the variance provides reasonable testing as long as the proportion is not close to zero or one. Its modification by Anscombe (1948), which replaces the observed proportion with $(Y + 3/8)/(n + 3/4)$, where Y is the observed number of successes out of n trials, was shown to provide even better variance stabilization.

A disadvantage of using transformations is that the choice of transformation is subjective. Mahmud et al. (2006) showed that trying several transformations on the data and then choosing one a posteriori inflates the type I error rate, making it much easier to falsely reject the null hypothesis. Hence, transformations can be problematic when a particular choice is not predetermined by other considerations.

A simple example of the advantage of generalized linear modeling with the appropriate distribution of the response variable occurs when the response data include zeroes. For example, consider an experiment in which the number of beneficial insects per square meter of field was measured. In the past, the common approach was to transform the observed count Y to $\log(Y + c)$, where $c > 0$, so as not to reduce the sample size by eliminating zero counts. The choice of c has been the subject of several studies (e.g., Hill, 1963; Draper and Hunter, 1969; Carroll, 1980) using methods ranging from a likelihood based approach (Box and Cox, 1964) to choosing the value of c that makes the distribution of the residuals from model fitting closest to normal. The problem is that unless performed correctly, this approach can introduce additional bias into the back-transformed values and should be avoided if not done carefully (Berry, 1987). Zeroes are usually not an issue for generalized linear models except in cases where the number of zeroes in the dataset is so large that the model parameters are near the limits of their possible values and the sample size is not sufficiently large, or where the data are actually a mixture of two distributions such as zero-inflated or zero-altered Poisson distributions (Lambert, 1992). In these cases, zero-altered (Min and Agresti, 2005) or zero-inflated models are extensions of generalized linear models.

Despite the stated disadvantages of transformations, one of the advantages of transformations lies in the use of normal theory linear models for which inferential statistics have known distributions. Except for the special case of the normal distribution, inference in generalized linear models depends on the asymptotic distribution of the statistic. The accuracy of the nominal confidence levels and error rates of tests depend on having sufficiently large sample sizes. For many distributions the sample size that would be considered sufficiently large will depend on the values of the parameters. As a simple example, the normal approximation to the binomial distribution when the success probability is close to one-half requires much smaller sample sizes than when the success probability is close to zero or one. There was a point in the development of generalized linear models when the statistics literature contained numerical studies for particular distributions and models but no comprehensive guidelines for what constitutes a sufficiently large sample size. That time is passing.

As stated in Chapter 1, generalized linear models have now entered the phase that statistical methodology goes through once high-quality comprehensive software to implement it becomes generally available. Normal theory mixed models went through this phase in the decade following the introduction of SAS's MIXED procedure in 1992. Similar maturation for generalized linear models was stimulated when GLIMMIX was introduced in 2005.

The question "How large a sample size is large enough?" is intimately tied to the question "Does my study design have adequate power to address my research objectives?" In Chapter 7, generalized linear mixed model based methods are presented to assess the power and precision of proposed designs whose primary response variable has a distribution belonging to the generalized linear mixed model family of distributions discussed in this text. As those methods are presented, simulation results to evaluate their accuracy are presented. Results to date suggest that issues with poor performance of generalized linear mixed models and their associated asymptotic statistics are strongly associated with under-powered studies. With adequately powered studies, the performance of generalized linear mixed model procedures and their associated inferential statistics has been uniformly encouraging. In addition, these studies suggest that transformations rarely compete well with generalized linear models for adequately powered studies. As will be seen in Chapter 5, this becomes even more emphatically the case when random effects complicate the modeling picture. Admittedly, there is more work to do, but following the planning aspect of research design and analysis using the methods shown in Chapter 7 is the best way to address the sufficiently large sample size question.

REFERENCES CITED

Agresti, A. 2002. Categorical data analysis. 2nd ed. John Wiley and Sons, Hoboken, NJ.

Akaike, H. 1974. A new look at the statistical model identification. IEEE Trans. Auto. Control AC 19:716–723. doi:10.1109/TAC.1974.1100705

Anscombe, F. 1948. The transformation of Poisson, binomial and negative binomial data. Biometrika 35:246–254.

Baker, R.J. 1980. Multiple comparison tests. Can. J. Plant Sci. 60:325–327. doi:10.4141/cjps80-053

Berry, D.A. 1987. Logarithmic transformations in ANOVA. Biometrics 43:439–456. doi:10.2307/2531826

Box, G.E.P., and D.R. Cox. 1964. An analysis of transformations. J. R. Stat. Soc. Ser. B (Methodological) 26:211–252.

Burnham, K.P., and D.R. Anderson. 2002. Model selection and multimodel inference: A practical information-theoretic approach. 2nd ed. Springer-Verlag, New York.

Bushong, J.T., T.L. Roberts, W.J. Ross, R.J. Norman, N.A. Slaton, and C.E. Wilson, Jr. 2008. Evaluation of distillation and diffusion techniques for estimating hydrolyzable amino sugar-nitrogen as a means of predicting nitrogen mineralization. Soil Sci. Soc. Am. J. 72:992–999. doi:10.2136/sssaj2006.0401

Carmer, S.G., and M.R. Swanson. 1973. An evaluation of ten pairwise multiple comparison procedures by Monte Carlo methods. J. Am. Stat. Assoc. 68:66–74. doi:10.2307/2284140

Carroll, R.J. 1980. A robust method for testing transformations to achieve approximate normality. J. R. Stat. Soc. Ser. B (Methodological) 42:71–78.

Chew, V. 1976. Comparing treatment means: A compendium. HortScience 11:348–357.

Curtiss, J.H. 1943. On transformations used in the analysis of variance. Ann. Math. Stat. 14:107–122. doi:10.1214/aoms/1177731452

Day, R.W., and G.P. Quinn. 1989. Comparisons of treatments after an analysis of variance in ecology. Ecol. Monogr. 59:433–483. doi:10.2307/1943075

Draper, N.R., and H. Smith. 1998. Applied regression analysis. 3rd ed. John Wiley and Sons, NY.

Draper, N.R., and W.G. Hunter. 1969. Transformations: Some examples revisited. Technometrics 11:23–40. doi:10.2307/1266762

Faraway, J.J. 2006. Extending the linear model with R: Generalized linear, mixed effects and nonparametric regression models. Chapman and Hall/CRC, Boca Raton, FL.

Fisher, R.A. 1921. On the "probable error" of a coefficient of correlation deduced from a small sample. Metron 1(4):3–32.

Fisher, R.A. 1954. The analysis of variance with various binomial transformations. Biometrics 10:130–151. doi:10.2307/3001667

Gill, J. 2001. Generalized linear models: A unified approach. Sage University Papers Series on Quantitative Applications in the Social Sciences, 07–134. Sage Publications, Thousand Oaks, CA.

Hill, B.M. 1963. The three-parameter lognormal distribution and Bayesian analysis of a point-source epidemic. J. Am. Stat. Assoc. 58:72–84. doi:10.2307/2282955

Hochberg, Y., and A.C. Tamhane. 1987. Multiple comparison procedures. John Wiley and Sons, NY.

Hurvich, C.M., and C.-L. Tsai. 1989. Regression and time series model selection in small samples. Biometrika 76:297–307. doi:10.1093/biomet/76.2.297

Kramer, C.Y. 1956. Extension of multiple range tests to group means with unequal numbers of replications. Biometrics 12:307–310. doi:10.2307/3001469

Lambert, D. 1992. Zero-inflated Poisson regression, with an application to defects in manufacturing. Technometrics 34:1–14. doi:10.2307/1269547

Mahmud, M., M. Abrahamowicz, K. Leffondré, and Y.P. Chaubey. 2006. Selecting the optimal transformation of a continuous covariate in Cox's regression: Implications for hypothesis testing. Commun. Stat. Simul. Comput. 35:27–45. doi:10.1080/03610910500415928

McArdle, B.H., and M.J. Anderson. 2004. Variance heterogeneity, transformations, and models of species abundance: A cautionary tale. Can. J. Fish. Aquat. Sci. 61:1294–1302. doi:10.1139/f04-051

McCullagh, P., and J.A. Nelder. 1989. Generalized linear models. 2nd ed. Chapman and Hall, NY.

Menard, S. 2002. Applied logistic regression analysis. 2nd ed. Quantitative Applications in the Social Sciences 106. Sage Publications, Thousand Oaks, CA.

Miller, R.G. 1981. Simultaneous statistical inference. 2nd ed. Springer-Verlag, NY.

Min, Y., and A. Agresti. 2005. Random effects models for repeated measures of zero-inflated count data. Stat. Model. 5:1–19. doi:10.1191/1471082X05st084oa

Oehlert, G.W. 1992. A note on the delta method. Am. Stat. 46:27–29. doi:10.2307/2684406

Ryan, G.W., and S. Leadbetter. 2002. On the misuse of confidence intervals for two means in testing for the significance of the difference between the means. J. Mod. Appl. Stat. Methods 1:473–478.

Scheffé, H. 1999. The analysis of variance. John Wiley and Sons, New York.

Schenker, N., and J.F. Gentleman. 2001. On judging the significance of differences by examining the overlap between confidence intervals. Am. Stat. 55:182–186. doi:10.1198/000313001317097960

Schwarz, G. 1978. Estimating the dimension of a model. Ann. Stat. 6:461–464. doi:10.1214/aos/1176344136

Winer, B.J., D.R. Brown, and K.M. Michels. 1991. Statistical principles in experimental design. 3rd ed. McGraw-Hill, New York.

LINEAR MIXED MODELS

4.1 INTRODUCTION

In this chapter normal theory linear mixed models are briefly reviewed. Several examples of designs and analyses commonly encountered in the agricultural sciences are presented. Issues that are important in generalized linear mixed models will be introduced in the context of this more familiar setting and revisited in Chapter 5.

Recall that a mixed model is one that contains at least one fixed and at least one random effect. Examples include designs with random blocks and split-plot type designs, as well as those having factors that are random effects. Although not explicitly mixed models, we also include linear models with correlated observations. The correlation structure can arise from repeated observations on the same experimental unit, spatial or temporal correlation, and random effects common to several experimental units.

A linear mixed model can be formulated as a conditional model or as a marginal model (Section 2.10). Usually it is presented in its conditional form, as it will be in the discussion to follow. However, the marginal form can be more useful in certain situations. This issue will be discussed further in Section 4.3 and will recur where relevant in subsequent examples in this chapter.

For a normally distributed response Y, let $x_1,..., x_p$ be a set of p explanatory variables describing the fixed effects and let $u_1,..., u_q$ be a set of q random effects. In conditional linear mixed models, the conditional mean for the jth observation given the random effects is expressed as

$$E\left[Y_j \mid u_1,..., u_q\right] = \beta_0 + \sum_{i=1}^{p} \beta_i x_{ij} + \sum_{k=1}^{q} z_{kj} u_k, j = 1,...,n$$

where β_i is the ith fixed effect coefficient, x_{ij} is the ith fixed effect explanatory variable for the jth observation, and z_{kj} is the binary indicator variable for the effect of the kth random effect, u_k, on the jth observation. In matrix form, the conditional mean for the mixed model can be written as

doi:10.2134/2012.generalized-linear-mixed-models.c4

Analysis of Generalized Linear Mixed Models in the Agricultural and Natural Resources Sciences
Edward E. Gbur, Walter W. Stroup, Kevin S. McCarter, Susan Durham, Linda J. Young, Mary Christman, Mark West, and Matthew Kramer

$$E[\mathbf{Y} \mid \mathbf{u}] = \mathbf{X}\boldsymbol{\beta} + \mathbf{Z}\mathbf{u}$$

where $\mathbf{Y}$ is the $n \times 1$ vector of responses, $\mathbf{Z}$ is the $n \times q$ design matrix for the random effects and $\mathbf{u}$ is the $q \times 1$ vector of random effects. As before, $\mathbf{X}$ is the $n \times (p+1)$ fixed effects design matrix and $\boldsymbol{\beta}$ is the $(p+1) \times 1$ vector of fixed effects coefficients. The conditional variance can be written as

$$\text{var}[\mathbf{Y} \mid \mathbf{u}] = \mathbf{R}$$

where $\mathbf{R}$ is an $n \times n$ covariance matrix. The conditional distribution of $\mathbf{Y}$ given $\mathbf{u}$ is multivariate normal, and the distribution of $\mathbf{u}$ is also multivariate normal with mean $\mathbf{0}$ and covariance matrix $\mathbf{G}$; i.e., $\mathbf{u} \sim \text{MVN}(\mathbf{0}, \mathbf{G})$. Here the symbol "$\sim$" is read as "is distributed as" or "has probability distribution."

4.2 ESTIMATION AND INFERENCE IN LINEAR MIXED MODELS

Inference in linear mixed models involves estimation and hypothesis testing related to the unknown parameters in $\boldsymbol{\beta}$, $\mathbf{R}$, and $\mathbf{G}$, as well as prediction of the random effects $\mathbf{u}$. Prediction will be discussed in Section 4.8.

For normally distributed responses $\mathbf{Y}$ and random effects $\mathbf{u}$, inference is based on approximate F statistics or t statistics that can be motivated from a Wald or likelihood ratio perspective (Section 2.6). REML is the standard approach for estimating the covariance matrices $\mathbf{G}$ and $\mathbf{R}$. Maximum likelihood yields the linear mixed model equations (Henderson, 1984). Solving these equations for $\mathbf{b}$ and $\mathbf{u}$ produces estimates in terms of the unknown $\mathbf{G}$ and $\mathbf{R}$. The solution $\hat{\boldsymbol{\beta}}$ is equivalent to the generalized least squares estimate (Searle, 1971). The ordinary least squares normal equations are a special case when the model contains no random effects and the observations are independent and homoscedastic.

For known $\mathbf{G}$ and $\mathbf{R}$, the solution $\hat{\boldsymbol{\beta}}$ is the best linear unbiased estimator (BLUE) of $\boldsymbol{\beta}$ and $\hat{\mathbf{u}}$ is the best linear unbiased predictor (BLUP) of $\mathbf{u}$ (Robinson, 1991; McLean et al., 1991). Here "best" is used in the sense of minimizing the mean squared error among all linear estimators or predictors, respectively. However, the mixed model equations for $\boldsymbol{\beta}$ and $\mathbf{u}$ usually must be solved in terms of the unknown $\mathbf{G}$ and $\mathbf{R}$. For unknown $\mathbf{G}$ and $\mathbf{R}$, their REML estimates can be substituted into the expressions for $\hat{\boldsymbol{\beta}}$ and $\hat{\mathbf{u}}$. The resulting estimators and predictors are often referred to as empirical BLUE (EBLUE) and empirical BLUP (EBLUP), respectively.

While inference involving $\boldsymbol{\beta}$ and $\mathbf{u}$ uses approximate t- and F-tests that can be motivated from a Wald perspective, the same is not true for the variance and covariance components. The Wald Z for these components is only valid for very large sample sizes that rarely occur in agricultural research. For this reason, its use is discouraged. The likelihood ratio test based on the χ^2 distribution is a better alternative. For confidence interval estimates on variance and covariance components, a profile likelihood approach can be used.

Mixed model methodology underwent extensive development in the 1980s and 1990s. Along with this development, researchers began to examine the behavior of mixed model procedures with the relatively small experiment sizes to which mixed model methods were likely to be applied. The two primary issues were degrees of freedom adjustments and small sample standard error bias. Giesbrecht and Burns (1985) developed a generalization of the Satterthwaite (1946) degree of freedom approximation applicable to linear mixed models in general. Kackar and Harville (1984) found that when estimated variance and covariance parameters are used to compute test statistics and interval estimates, the resulting test statistics tend to be biased upward and the standard errors used to compute confidence intervals tend to be biased downward. This problem does not occur in balanced data sets without covariance components. It occurs, but only to a negligible degree, in models for unbalanced data sets without covariance components such as complete block designs with missing data and incomplete block designs. However, for models with covariance components, whether balanced or not, bias occurs, and the more complex the model, the more severe the bias. Kenward and Roger (1997) derived a bias correction term that was easily implemented as part of mixed model analysis procedures. They also refined Giesbrecht and Burns' degree of freedom approximation for use with correlated error models. These are collectively known as the Kenward–Roger adjustment and should be considered standard operating procedure for linear mixed models.

4.3 CONDITIONAL AND MARGINAL MODELS

Conditional and marginal models were introduced in Section 2.10. In this section, the distinction between these models is developed in the context of normal theory linear mixed models. Using the notation of Section 4.1, the essential elements of the conditional linear mixed model can be summarized in matrix form as follows.

- The conditional distribution of the response given the random effects is $Y \mid u \sim \text{MVN}(E[Y \mid u], R)$.
- The distribution of the random effects is $u \sim \text{MVN}(0, G)$.
- The link function is the identity function.
- The linear predictor is $\eta = E[Y \mid u] = X\beta + Zu$.

If we re-express the conditional model without explicit reference to the random effects, we obtain the marginal model.

- The distribution of the response is $Y \sim \text{MVN}(E[Y], V)$, where $V = ZGZ' + R$.
- The linear predictor is $\eta = E[Y] = X\beta$.

The marginal model excludes the random effects from the linear predictor but includes all of their variance information in the variance–covariance structure of the response. When all of the probability distributions are normal, the marginal and conditional models result in identical estimates of β and identical inference for the fixed effects. Example 4.1 will demonstrate this equivalence. Example 4.2 will provide insight as to why it is useful. As we will see in Chapter 5, this conditional–marginal model equivalence holds only for the normal distribution.

As shown above, the random effects in a mixed model can be contained in either **G** or **R** or both and are labeled as G-side and R-side effects, respectively. Models fit with GLIMMIX may have G-side effects, R-side effects, both, or neither. Marginal models have no G-side effects. R-side effects are also referred to as residual effects.

EXAMPLE 4.1

In this example we return to the randomized complete block design with ten blocks and two treatments in Example 2.10. Data have been constructed specifically to demonstrate the relationships between the conditional and marginal model formulations. The data are shown in Table 4–1. For this example, assume that the variable Y_1 is normally distributed. The model equation for the conditional model under the normality assumption can be written

$$Y_{ij} = \mu + T_i + b_j + w_{ij}$$

where μ is the intercept, T_i is the ith treatment effect, b_j is the jth block effect, and w_{ij} is the residual associated with the observation on the ith treatment in the jth block.

TABLE 4.1. Constructed data for Examples 4.1 and 4.2.

Block	Treatment	F	N	Y_1	Y_2
1	0	86	100	21.3	39.5
1	1	98	100	10.0	30.3
2	0	48	100	19.7	32.6
2	1	93	100	22.0	33.6
3	0	87	100	20.2	32.1
3	1	43	100	17.6	38.8
4	0	64	100	20.0	41.9
4	1	89	100	16.0	35.1
5	0	99	100	17.4	29.1
5	1	100	100	20.3	34.1
6	0	52	100	26.0	42.9
6	1	49	100	25.6	31.8
7	0	89	100	19.7	40.8
7	1	96	100	16.2	28.2
8	0	63	100	18.5	40.9
8	1	98	100	12.4	38.1
9	0	48	100	20.9	39.5
9	1	83	100	13.5	34.7
10	0	85	100	23.0	42.0
10	1	97	100	23.2	32.8

Since b_j and w_{ij} are independent and normally distributed, their sum, $e_{ij} = b_j + w_{ij}$ is also normally distributed. The model equation for the marginal model can be rewritten as

$$Y_{ij} = \mu + T_i + e_{ij}$$

Expressing this in terms of the essential elements of a linear model, we have

- The distribution of the response is $\mathbf{Y} \sim \text{MVN}(E[\mathbf{Y}], \mathbf{V})$.
- The linear predictor is $\eta_{ij} = E[Y_{ij}] = \mu + T_i$.

The structure of $\mathbf{V}$ follows from the variance and covariance of e_{ij}. Specifically,

- For any observation, $\sigma_e^2 = \text{var}(e_{ij}) = \text{var}(b_j + w_{ij}) = \sigma_B^2 + \sigma_W^2$.
- For two observations from the same block, $\text{cov}(e_{ij}, e_{ij'}) = \sigma_B^2 = \rho\sigma_e^2$, where $\rho = \sigma_B^2 / (\sigma_B^2 + \sigma_W^2)$.
- For two observations from different blocks, $\text{cov}(e_{ij}, e_{i'j'}) = 0$.

This covariance structure is referred to as compound symmetry (CS). The correlation ρ appears in many contexts where it is called the intraclass correlation coefficient. Note the two defining features of this marginal model.

- There are no random effects in the linear predictor.
- The block variance is contained in the covariance structure.

Figure 4–1 shows the GLIMMIX statements for the conditional and marginal models. The default distribution used in GLIMMIX is the normal so the *link* and *dist* options are not required on the *model* statement. For the conditional model, the *block* random effect is specified using the *random* statement. For the marginal model, the *random _residual_* statement is GLIMMIX syntax for specifying the covariance structure of the response variable for something other than independent,

FIG. 4–1. GLIMMIX statements to fit the conditional and marginal models for Example 4.1.

```
title 'CONDITIONAL MODEL' ;
proc glimmix data=CvsM;
   class trt block;
   model Y1 = trt;
   random intercept / subject=block;
run;

title 'MARGINAL MODEL' ;
proc glimmix data=CvsM;
   class trt block;
   model Y1 = trt;
   random _residual_ / type=cs subject=block V;
run;
```

FIG. 4-2. GLIMMIX output containing estimated variance components and test for the fixed effects for the conditional model in Example 4.1.

Covariance Parameter Estimates

Cov Parm	Subject	Estimate	Standard Error
Intercept	block	5.6193	5.5187
Residual		9.9538	4.6923

Type III Tests of Fixed Effects

Effect	Num DF	Den DF	F Value	Pr > F
trt	1	9	4.49	0.0631

homoscedastic observations; i.e., when all variances are equal and all covariances are zero. For the *type* = *cs* model, the covariance within a block is not zero but is the same for all blocks.

Figure 4–2 shows the variance component estimates and tests of the fixed effects for the conditional model. The variance component estimates are $\hat{\sigma}_B^2 = 5.62$ and $\hat{\sigma}_W^2 = 9.95$ and the F-statistic for testing the null hypothesis $H_0: T_1 = T_2 = 0$ (i.e., equal treatment means or equivalently, no treatment effect) is $F = 4.49$ with a p-value of 0.0631.

Figure 4–3 shows the analogous results for the marginal model. In addition, the *Estimated V Matrix for block* 1 part of the output contains the portion of the covariance matrix $\mathbf{V}$ that corresponds to the first observation. The diagonal elements represent the estimate var(e_{ij}) = $\hat{\sigma}_e^2 = 15.57$, which equals the sum of $\hat{\sigma}_B^2$ and $\hat{\sigma}_W^2$ from the conditional model, as theory says it should. The off-diagonal elements are estimates of cov($e_{ij}, e_{ij'}$) = 5.62, which, as theory says, equals $\hat{\sigma}_B^2$ from the conditional model.

Note how the compound symmetry covariance component estimates are labeled. Both SAS's (SAS Institute, Cary, NC) MIXED and GLIMMIX procedures use this format to underscore the conditional-marginal model equivalence. Finally, observe that the F-statistics are identical for the two model fits. Estimates of treatment means, standard errors, interval estimates, and hypothesis tests are identical for both models. ■

PROGRAMMING NOTE: The *random* statement in Fig. 4–1 for the conditional model can be expressed equivalently by the following simpler alternative statement.

random block;

When the form shown in Fig. 4–1 is used, the entries of **u** are sorted so that all terms for a block are grouped together and the columns of **Z** are rearranged accordingly. This rearrangement allows GLIMMIX to process the data more efficiently. In addition, as will be seen later, the form in Fig. 4–1 becomes important

FIG. 4-3. GLIMMIX output containing the entries of the matrix **V** for the first block, the estimated variance components and test for the fixed effects for the marginal model in Example 4.2.

Estimated V Matrix for block 1		
Row	Col1	Col2
1	15.5732	5.6193
2	5.6193	15.5732

Covariance Parameter Estimates			
Cov Parm	Subject	Estimate	Standard Error
CS	block	5.6193	5.5187
Residual		9.9538	4.6923

Type III Tests of Fixed Effects				
Effect	Num DF	Den DF	F Value	Pr > F
trt	1	9	4.49	0.0631

for models requiring more than one *random* statement and when the distribution of the response is non-normal.

In addition, for the marginal model in Fig. 4–1, the *random* statement can be expressed in the alternative form

random variety / subject = block type = cs residual v;

EXAMPLE 4.2

This example is a continuation of Example 4.1 using the response variable denoted Y_2. For this example, Y_2 will be assumed to be normally distributed. The same GLIMMIX statements in Fig. 4–1 were used to fit the models with Y_2 replacing Y_1. Figures 4–4 and 4–5 give the conditional and marginal model results, respectively. Note that the results are not the same. For the conditional model, $\hat{\sigma}_B^2 = 0, \hat{\sigma}_W^2 = 17.43$, $F = 5.50$, and $p = 0.0436$. For the marginal model, from the **V** matrix results, the estimated variance is $\text{var}(e_{ij}) = 17.43$, which equals the conditional $\hat{\sigma}_B^2 + \hat{\sigma}_W^2$ as it should, but the estimated covariance is $\text{cov}(e_{ij}, e_{ij'}) = -5.21$, clearly not equal to $\hat{\sigma}_B^2$. In addition, for the marginal model, $F = 4.24$ with $p = 0.0696$. Why is there a discrepancy? Since variety effects appear to be significant under the conditional model (assuming $\alpha = 0.05$ as the criterion) but not under the marginal model, which result would be correct to report?

The discrepancy results from the REML estimate of the block variance σ_B^2 in the conditional model. The REML estimating equations yielded a negative solution

FIG. 4-4. GLIMMIX output containing estimated variance components and test for the fixed effects for the conditional model in Example 4.2.

Covariance Parameter Estimates

Cov Parm	Subject	Estimate	Standard Error
Intercept	block	0	.
Residual		17.4270	5.8090

Type III Tests of Fixed Effects

Effect	Num DF	Den DF	F Value	Pr > F
trt	1	9	5.50	0.0436

FIG. 4-5. GLIMMIX output containing the entries of the matrix **V** for the first block, the estimated variance components and test for the fixed effects for the marginal model in Example 4.2.

Estimated V Matrix for block 1

Row	Col1	Col2
1	17.4270	-5.2072
2	-5.2072	17.4270

Covariance Parameter Estimates

Cov Parm	Subject	Estimate	Standard Error
CS	block	-5.2072	6.0628
Residual		22.6342	10.6699

Type III Tests of Fixed Effects

Effect	Num DF	Den DF	F Value	Pr > F
trt	1	9	4.24	0.0696

for $\hat{\sigma}_B^2$ but since variance by definition must be non-negative, and a literal reading of likelihood theory says that estimators cannot lie outside the parameter space (i.e., negative variances are impossible), $\hat{\sigma}_B^2$ is set to zero. If one reverts to pre-mixed model days and computes an ANOVA table, $F = \text{MS(Variety)}/\text{MS(Error)} = 95.92/22.63 = 4.24$. Note that this is identical to the result for the marginal model.

Stroup and Littell (2002) investigated the consequences of setting negative estimates to zero in the conditional model and found that it is generally not a good idea to do so if the primary objective is to obtain accurate inference on

FIG. 4-6. GLIMMIX output containing estimated variance components and tests for the fixed effects for the conditional model in Example 4.2 with the *nobound* option added.

Covariance Parameter Estimates

Cov Parm	Subject	Estimate	Standard Error
Intercept	block	−5.2072	6.0628
Residual		22.6342	10.6699

Type III Tests of Fixed Effects

Effect	Num DF	Den DF	F Value	Pr > F
trt	1	9	4.24	0.0696

treatment effects. In a block design, negative $\hat{\sigma}_B^2$ occurs whenever MS(block) < MS(error), which can occur as a consequence of the sampling distribution of the mean squares even when σ_B^2 is substantially greater than zero. Setting $\hat{\sigma}_B^2$ to zero biases the block variance estimate upward, which in turn, biases the error variance downward, which in turn, biases F-statistics upward. For this reason, the general recommendation for linear mixed models is to allow the variance component estimates to remain negative and not set them to zero. This may be hard to explain when reporting the variance component estimate per se, but it is essential if accurate inference on treatment effects is considered the greater good. Note that this issue does not arise with the marginal model because the variability among blocks is modeled as a covariance, and a negative covariance is well-defined.

For the conditional model, adding the *nobound* option to the PROC GLIMMIX statement overrides the set-to-zero default and allows the variance estimates to remain negative. The *nobound* option would have no effect on the marginal model since the parameter space for covariance is not bounded at zero. The results for the conditional model with the *nobound* option are shown in Fig. 4–6. The variance component estimates are now $\hat{\sigma}_B^2 = -5.21, \hat{\sigma}_W^2 = 22.63, F = 4.24$, and $p = 0.0696$, all of which now agree with the marginal model (and with the ANOVA F-test). ∎

4.4 SPLIT PLOT EXPERIMENTS

Split plot designs are one of the most commonly used designs in field experimentation. They occur when experimental units for some factors are a different size than the experimental units for other factors. For example, in a field crop experiment to study the effect of irrigation timing on different cultivars, it would be impractical to assign different irrigation timings to each plot separately. Plots could be grouped and timing assigned randomly to the groups of plots. Each plot within a group would be assigned randomly to a cultivar. The set of plots would constitute the experimental unit for the irrigation factor (whole plot), and an individual plot would be the experimental unit for the cultivar factor (split plot). It is important to use different randomizations of the split plot experimental units within each whole plot experimental unit. A common mistake that is made in this type of experiment is the failure to replicate the whole plot treatments. In this example, this would occur if only one set of plots had been assigned to each irrigation treatment, even if the cultivars were replicated within the set of irrigation treatment plots.

Split plot designs are also applicable to laboratory experiments. For example, in a biodegradation experiment, soil samples contaminated with hexadecane were incubated at one of two temperatures. Half of the samples had nitrogen added, and the other half did not. The response was the amount of nitrate nitrogen recovered in a fixed period of time. Nitrate nitrogen levels provide an indirect measure of microbial degradation of the hexadecane. The entire experiment was run a total of three times. Since different temperatures required the use of different incubators, temperature was the whole plot factor. The runs of the experiment formed the blocks of a randomized complete block for the whole plot portion of the design. The split plot factor was the addition of nitrogen.

EXAMPLE 4.3

As part of a study of the response of early maturing soybean to increased leaflet number, Seversike et al. (2009) compared the mean leaf area per leaf for four pairs of 3- and 7-leaflet near-isogenic lines at the R5 developmental stage. The design of the field experiment was a split plot where the whole plot structure was a randomized complete block with four blocks and four genotypes. The split plot factor was leaflet number (3 or 7). Genotype and leaflet number were fixed effects, and blocks were random. One observation was excluded from the analysis as an outlier. The remaining 31 observations were included in the analysis.

Since leaf area was assumed to be normally distributed, the link function is the identity function $\eta = g(\mu) = \mu$. The mean leaf area per leaf, Y, conditional on the observed random effects, is given by

$$E[Y_{ijk} \mid B_k, w_{ik}] = \beta_0 + G_i + L_j + GL_{ij} + B_k + w_{ik} \text{ for } i = 1, 2, 3, 4; j = 1, 2; k = 1, 2, 3, 4,$$

where β_0 is the intercept, G_i is the genotype (near-isogenic line) effect, L_j is the leaflet number effect, GL_{ij} is the genotype × leaflet number interaction effect, B_k is the block effect, and w_{ik} is the whole plot error effect. Note that since we are conditioning on the random effects, their observed values are used in the expression for the conditional mean.

The model can be written in matrix form as

$$E[\mathbf{Y} \mid \mathbf{u}] = g^{-1}(\mathbf{X}\beta + \mathbf{Zu}) = \mathbf{X}\beta + \mathbf{Zu}$$

where the random effects are multivariate normal with $\mathbf{u} \sim MVN(\mathbf{0}, \mathbf{G})$. The conditional variance of Y is given by

$$var[\mathbf{Y} \mid \mathbf{u}] = \mathbf{R}$$

FIG. 4-7. GLIMMIX statements to fit the model for Example 4.3.

```
proc glimmix data=t6 plots=studentpanel(BLUP);
  class block geno leaflet;
  model Area = geno leaflet geno*leaflet / ddfm=kr;
  random block block*geno;
  lsmeans geno*leaflet / plot=(diffplot(noabs)) plot=meanplot(sliceby=leaflet join cl);
  lsmeans geno*leaflet / slice=geno slicediff=geno adjust=Tukey;
  lsmeans geno*leaflet / slice=leaflet slicediff=leaflet adjust=Tukey;
  output out=new pred=pred stderr=sepred resid=resid student=student;
quit;
```

The vector β contains the 15 fixed effects parameters (overall mean + 4 genotype effects + 2 leaflet number effects + 8 interaction effects) and the vector **u** contains the 20 random effects (4 block effects + 16 whole plot error terms). **X** is the 31 × 15 fixed effects design matrix, and **Z** is the 31 × 20 random effects design matrix. There are two parameters, the block variance and the whole plot error variance, that appear in the random effects covariance matrix **G** and one parameter, the residual or split plot error variance, that appears in the residual covariance matrix **R**. **G** is a diagonal matrix with the block variance as the first four entries on the diagonal, followed by the whole plot error variance. **R** is also a diagonal matrix with the residual or split plot error variance on the diagonal.

The GLIMMIX statements used to fit this model are given in Fig. 4–7. The mean leaf area per leaf is denoted by *area* and the genotype effect by *geno*. Normality and the identity link are the defaults in GLIMMIX and do not need to be specified as options on the *model* statement. The Kenward–Roger adjustment to the degrees of freedom was requested on the *model* statement. The random effects are specified on the *random* statement where *block*geno* represents whole plot error. The equivalent formulation

random intercept geno/subject = block;

could have been used for the *random* statement.

Figure 4–8 contains the basic model information. The *Model Information* section lists the response variable, its distribution, and link function. The default estimation method is residual maximum likelihood (REML). In contrast to PROC MIXED, the *Dimensions* section lists the number of G-side and R-side covariance parameters separately. The figure indicates that there are two G-side parameters. The block effect has four levels, but only a single variance component is estimated for block. Similarly the block × genotype effect has 16 levels but only a single covariance parameter. Because a *subject* option was not included on the *random* statement, GLIMMIX treats these data as having come from a single subject.

The variance component estimates are given in Fig. 4–9 along with estimated standard errors. In this example, the output contained the message "*Convergence*

FIG. 4-8. GLIMMIX output containing basic model and fitting information for Example 4.3.

Model Information

Data Set	WORK.T6
Response Variable	Area
Response Distribution	Gaussian
Link Function	Identity
Variance Function	Default
Variance Matrix	Not blocked
Estimation Technique	Restricted Maximum Likelihood
Degrees of Freedom Method	Kenward–Roger
Fixed Effects SE Adjustment	Kenward–Roger

Class Level Information

Class	Levels	Values
block	4	1 2 3 4
geno	4	Hendrick Mn1401 Mn1801 Traill
leaflet	2	3 7

Dimensions

G-side Cov. Parameters	2
R-side Cov. Parameters	1
Columns in X	15
Columns in Z	20
Subjects (Blocks in V)	1
Max Obs per Subject	31

FIG. 4-9. GLIMMIX output containing estimated variance components for Example 4.3.

Covariance Parameter Estimates

Cov Parm	Estimate	Standard Error
block	0	.
block*geno	278.42	130.95
Residual	78.4679	33.2092

criterion (GCONV = 1E-8) satisfied. Estimated G matrix is not positive definite." In this case, the nonpositive definiteness of the random effects covariance matrix **G** arises because the estimated block variance is zero. This is the split plot version of the problem with the block variance in Example 4.2.

FIG. 4-10. GLIMMIX output containing estimated variance components using the *nobound* option to allow variance estimates to be negative in Example 4.3.

Covariance Parameter Estimates		
Cov Parm	Estimate	Standard Error
block	−69.0692	50.7166
block*geno	346.64	182.74
Residual	78.5032	33.2329

Examination of the data found that the block means ranged from 82.5 to 94.3, indicating some amount of block variability. However, the variability among the blocks was less than the background variability within blocks. In classical analysis of variance terminology, the negative variance estimate occurs because MS(Blocks) < MS(Whole Plot Error). This may be the result of blocks that were formed for convenience rather than from application of a criterion that corresponded to actual differences among the whole plot experiment units. Following Stroup and Littell (2002) as in Example 4.2, the model should be expanded to allow negative variance component estimates by using either the *nobound* option or the equivalent compound symmetry covariance structure.

The model was refit by adding the *nobound* option to the PROC GLIMMIX statement in Fig. 4–7. The variance estimates from the fit are given in Fig. 4–10. The block variance estimate was negative, and the whole plot variance estimate increased, while the split plot variance estimate was essentially unchanged. ∎

As shown in Example 4.1, in a randomized complete block design the parameter representing the block variance also represents the covariance between observations from experimental units in the same block. Hence, setting negative estimates to zero implicitly assumes that the covariance must be non-negative. There is no theoretical reason that the correlation must be positive, and in some applications a negative correlation might be a reasonable assumption. For example, in experiments where there may be competition for resources among experimental units in the same block, negative correlations could result. When a common covariance is assumed for all pairs of observations within the same block, the covariance structure is called compound symmetry.

In a split plot where the whole plot portion is a randomized complete block design, the covariance structure is slightly more complicated. The covariance between split plot experimental units in the same whole plot experimental unit (and hence, in the same block) is the sum of the block and whole plot error variances while the covariance between different whole plot experimental units in the same block is still the block variance. Within each block, the covariance structure can still be represented by compound symmetry. Generally observations from different blocks are assumed to be independent.

Based on the above, the entries in the covariance matrix $\mathbf{V}$ of the response $\mathbf{Y}$ are as follows, where e represents the split plot error and σ_B^2 is interpreted as the block variance or the covariance between whole plot experimental units in the same block.

- For any observation, $\text{var}\left(Y_{ijk} \right) = \sigma_B^2 + \sigma_w^2 + \sigma_e^2$.
- For any pair of observations in the same whole plot experimental unit (and hence, the same block), $\text{cov}\left(Y_{ijk}, Y_{ijk'} \right) = \sigma_B^2 + \sigma_w^2$.
- For any pair of observations in the same block but different whole plot experimental units, $\text{cov}\left(Y_{ijk}, Y_{ij'k'} \right) = \sigma_B^2$.
- For any pair of observations in different blocks, $\text{cov}\left(Y_{ijk}, Y_{i'j'k'} \right) = 0$.

EXAMPLE 4.4

This example is a continuation of Example 4.3 that explicitly recognizes the within block compound symmetry covariance structure. To fit this model using GLIMMIX the *random* statement in Fig. 4–7 is replaced by

random geno / subject = block type = cs g v;

The *type* option defines the covariance structure for the genotypes (whole plot factor levels) within a block. Since the block variance is being represented by the covariance in the compound symmetry structure, the *intercept* term that represents the block variance in the *subject* option form of the *random* statement is not included. The *g* and *v* options have been added to produce portions of the **G** and **V** matrices as part of the output for illustrative purposes. They are not necessary for modeling the covariance structure. In addition, by default, the *cs* option allows negative variance estimates and hence, the *nobound* option on the *proc* statement is no longer necessary. The remaining statements in Fig. 4–7 are unchanged.

Figure 4–11 contains the basic model information. Compared to Fig. 4–8 for the fit without an assumption about the covariance structure, the difference is the

FIG. 4–11. GLIMMIX output containing basic model and fitting information assuming a compound symmetry covariance structure for Example 4.4.

Model Information	
Data Set	WORK.T6
Response Variable	Area
Response Distribution	Gaussian
Link Function	Identity
Variance Function	Default
Variance Matrix Blocked By	block
Estimation Technique	Restricted Maximum Likelihood
Degrees of Freedom Method	Kenward–Roger
Fixed Effects SE Adjustment	Kenward–Roger

Dimensions	
G-side Cov. Parameters	2
R-side Cov. Parameters	1
Columns in X	15
Columns in Z per Subject	4
Subjects (Blocks in V)	4
Max Obs per Subject	8

FIG. 4–12. GLIMMIX output containing the estimated covariance parameters assuming a compound symmetry covariance structure for Example 4.4.

Covariance Parameter Estimates			
Cov Parm	Subject	Estimate	Standard Error
Variance	block	346.64	182.74
CS	block	−69.0692	50.7166
Residual		78.5032	33.2329

FIG. 4–13. GLIMMIX output containing the portion of the estimated **G** matrix for the covariance structure within the first block assuming a compound symmetry covariance structure for Example 4.4.

Estimated G Matrix						
Effect	Genotype	Row	Col1	Col2	Col3	Col4
geno	Hendrick	1	277.57	−69.0692	−69.0692	−69.0692
geno	Mn1401	2	−69.0692	277.57	−69.0692	−69.0692
geno	Mn1801	3	−69.0692	−69.0692	277.57	−69.0692
geno	Traill	4	−69.0692	−69.0692	−69.0692	277.57

number of columns in **Z** per subject (block). In the current fit, the four columns per subject in **Z** and the corresponding block in **G** represent the covariance structure among the genotypes within a block. **G** does not directly include an estimate of the variance among whole plot experimental units.

The covariance parameter estimates in Fig. 4–12 are a rearrangement and relabeling of the estimates in Fig. 4–10. The whole plot error variance estimate (346.64) is reported in the first row, the covariance between whole plot experimental units in the same block (or equivalently, the block variance) in the second row (−69.07), and the residual or split plot variance in the last row.

The portion of the **G** matrix for the covariance structure within the first block is shown in Fig. 4–13. The diagonal entries (277.57 = −69.07 + 346.64) represent the covariance between 3-and 7-leaflet plants of the same genotype (i.e., the covariance between split plot experimental units in the same whole plot experimental unit) and the off-diagonal entries (−69.07) represent the covariance between observations on different genotypes regardless of leaflet number (i.e., the covariance between different whole plot experimental units in the same block).

The portion of the estimated covariance matrix **V** for the response (mean leaf area per leaf) for the first block is shown in Fig. 4–14. Each 2 × 2 block on the diagonal beginning with the first row and column corresponds to a whole plot experimental unit, and the rows and columns within the 2 × 2 block correspond to the two split plot experimental units in that whole plot experimental unit. The diagonal entries (356.07 = −69.07 + 346.64 + 78.50) are the variances of the response.

FIG. 4–14. GLIMMIX output containing the portion of the estimated covariance matrix **V** for the response (mean leaf area per leaf) within the first block assuming a compound symmetry covariance structure for Example 4.4.

Estimated V Matrix for block 1							
Row	Col1	Col2	Col3	Col4	Col5	Col6	Col7	Col8
1	356.07	277.57	−69.0692	−69.0692	−69.0692	−69.0692	−69.0692	−69.0692
2	277.57	356.07	−69.0692	−69.0692	−69.0692	−69.0692	−69.0692	−69.0692
3	−69.0692	−69.0692	356.07	277.57	−69.0692	−69.0692	−69.0692	−69.0692
4	−69.0692	−69.0692	277.57	356.07	−69.0692	−69.0692	−69.0692	−69.0692
5	−69.0692	−69.0692	−69.0692	−69.0692	356.07	277.57	−69.0692	−69.0692
6	−69.0692	−69.0692	−69.0692	−69.0692	277.57	356.07	−69.0692	−69.0692
7	−69.0692	−69.0692	−69.0692	−69.0692	−69.0692	−69.0692	356.07	277.57
8	−69.0692	−69.0692	−69.0692	−69.0692	−69.0692	−69.0692	277.57	356.07

FIG. 4–15. GLIMMIX output containing the results of the tests for the fixed effects for Example 4.4.

Type III Tests of Fixed Effects				
Effect	Num DF	Den DF	F Value	Pr > F
geno	3	9.107	6.29	0.0134
leaflet	1	11.37	19.63	0.0009
geno*leaflet	3	11.35	4.10	0.0340

The entries not in the 2 × 2 diagonal blocks correspond to the covariance between different whole plot experimental units (genotypes) in the same block.

Based on the F-tests for fixed effects in Fig. 4–15, there was a significant genotype × leaflet number interaction. Hence, subsequent analyses should involve only the cell means (simple effects) and not the main effects of genotype and leaflet number. By default, the means are computed on the link scale and not on the original data scale. Since the mean leaf area was assumed to be normally distributed, the link function is the identity function and back-transformation is not necessary in this example.

The first *lsmeans* statement in Fig. 4–7 produces the estimated least squares means for each genotype–leaflet number combination in tabular form in Fig. 4–16 and graphically in Fig. 4–17, the latter as a result of the *meanplot* option. The form of the interaction is clearly evident in the plot.

All possible estimated pairwise comparisons can also be summarized graphically by the *diffplot* option (Fig. 4–18). In this figure, known as a diffogram, the horizontal and vertical axes represent the least squares means. For each comparison, the coordinates of the intersection point of the two solid pale gray grid lines within the plot are the least squares means of the treatment combination identified by the labels on the grid lines. The confidence interval for the pairwise difference

FIG. 4-16. GLIMMIX output containing the estimated genotype × leaflet number means, estimated standard errors and tests of the hypothesis that the mean is zero for Example 4.4.

geno*leaflet Least Squares Means

Genotype	Leaflet number	Estimate	Standard Error	DF	t Value	Pr > |t|
Hendrick	3	77.2250	9.4350	13.03	8.18	<0.0001
Hendrick	7	71.0712	10.1055	15.45	7.03	<0.0001
Mn1401	3	85.9925	9.4350	13.03	9.11	<0.0001
Mn1401	7	109.22	9.4350	13.03	11.58	<0.0001
Mn1801	3	107.81	9.4350	13.03	11.43	<0.0001
Mn1801	7	131.72	9.4350	13.03	13.96	<0.0001
Traill	3	56.3850	9.4350	13.03	5.98	<0.0001
Traill	7	73.1875	9.4350	13.03	7.76	<0.0001

FIG. 4-17. GLIMMIX output displaying the genotype × leaflet number means and 95% confidence intervals for Example 4.4.

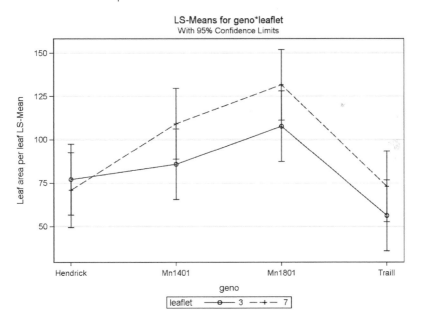

is represented by the line with negative slope centered at the associated intersection. Based on the equivalence of confidence intervals and hypothesis tests for comparing two means, the least squares means are not significantly different if the confidence interval line for the difference intersects the dashed 1–1 diagonal line representing equal means. The use of color, when available, and dashed versus solid confidence interval lines simplifies the interpretation of the results. In applications where it is not of interest to compare all possible pairs of means or when

FIG. 4–18. GLIMMIX output displaying 95% confidence intervals for differences between least squares means for all possible pairs of genotype–leaflet number combinations for Example 4.4.

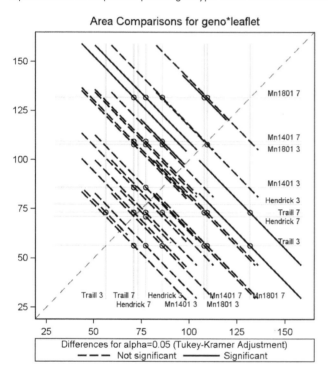

the number of pairwise comparisons is large, the *diffplot* does not provide much useful information.

For example, in Fig. 4–18, genotype Mn1801 with 7 leaflets is significantly different than Traill with 7 leaflets (leftmost solid line centered on the Mn1801-7 horizontal grid line) but is not significantly different than Mn1401 with 7 leaflets (rightmost dashed line centered on the Mn1801-7 horizontal grid line).

The *slice* and *slicediff* options in the second and third *lsmeans* statements in Fig. 4–7 produce the tests and mean comparisons for the simple effects by genotype and leaflet number, respectively. The tests are shown in Fig. 4–19 and the mean comparisons in Fig. 4–20. Note that the adjusted *p*-values are based on the number of comparisons within a group and not on the total number of comparisons. ∎

Three important facts about split plot designs emerge from this series of examples. First, correlation among observations is implicit in the split plot model even when all random effects and error terms are assumed to be mutually independent. Second, the covariance structure arising from the independence assumptions is compound symmetry if there is blocking in the whole plot portion of the design. Third, if compound symmetry does not adequately reflect the nature of the correlations among the responses, an alternative covariance structure should be explicitly incorporated into the model.

FIG. 4-19. GLIMMIX output containing F-tests for simple effects for Example 4.4.

Tests of Effect Slices for geno*leaflet Sliced By geno

Genotype	Num DF	Den DF	F Value	Pr > F
Hendrick	1	11.88	0.72	0.4119
Mn1401	1	11.16	13.75	0.0034
Mn1801	1	11.16	14.56	0.0028
Traill	1	11.16	7.19	0.0211

Tests of Effect Slices for geno*leaflet Sliced By leaflet

Leaflet number	Num DF	Den DF	F Value	Pr > F
3	3	10.81	4.27	0.0321
7	3	11.26	7.90	0.0041

4.5 EXPERIMENTS INVOLVING REPEATED MEASURES

Repeated measures experiments, also called longitudinal studies, are used to study changes over time (e.g., over a growing season) or space (e.g., over depths in soil core sampling) and the effect of treatments on these changes. Formally, a repeated measures experiment is defined as an experiment in which the experimental units are observed at two or more planned times or places over the course of the experiment. Repeated measures experiments can be conducted in conjunction with any design. They should not be confused with designs having multiple measurements taken on the same experimental unit at essentially the same time in the same place. The key feature of repeated measures is the objective of measuring changes in the response over a relevant interval of time or space.

There are two main issues in repeated measures analysis: namely, constructing a suitable model for the mean and selecting an adequate but parsimonious model for the covariance structure of the repeated measure. The model for the mean is determined by the design of the experiment, ignoring the repeated measurements aspect. The experimental unit on which the measurements will be repeated through time or space is referred to as the subject. The model for the portion of the experiment that does not include the repeated measure is called the between subjects model and the portion of the model that involves the repeated measurements is called the within subjects model. A fundamental premise of repeated measures modeling is that observations on the same subject are likely to be correlated. Moreover, observations that are closer together in time or space tend to be more highly correlated than observations farther apart in time or space. This type of correlation is known formally in repeated measures theory as serial correlation.

Superficially, the analysis of repeated measures experiments resembles that of split plot experiments with time as the split plot factor. The primary difference between the split plot and repeated measures models rests with their assumptions

FIG. 4–20. GLIMMIX output containing comparisons of the simple effects for Example 4.4.

Simple Effect Comparisons of geno*leaflet Least Squares Means By geno

Simple Effect Level	Leaflet number	Leaflet number	Estimate	Standard Error	DF	t Value	Pr > |t|	Adj P
geno Hendrick	3	7	6.1538	7.2356	11.88	0.85	0.4119	0.4126
geno Mn1401	3	7	−23.2300	6.2651	11.16	−3.71	0.0034	0.0033
geno Mn1801	3	7	−23.9075	6.2651	11.16	−3.82	0.0028	0.0027
geno Traill	3	7	−16.8025	6.2651	11.16	−2.68	0.0211	0.0208

Simple Effect Comparisons of geno*leaflet Least Squares Means By leaflet

Simple Effect Level	Genotype	Genotype	Estimate	Standard Error	DF	t Value	Pr > |t|	Adj P
leaflet 3	Hendrick	Mn1401	−8.7675	14.5798	10.81	−0.60	0.5600	0.9296
leaflet 3	Hendrick	Mn1801	−30.5850	14.5798	10.81	−2.10	0.0603	0.2114
leaflet 3	Hendrick	Traill	20.8400	14.5798	10.81	1.43	0.1811	0.5074
leaflet 3	Mn1401	Mn1801	−21.8175	14.5798	10.81	−1.50	0.1631	0.4706
leaflet 3	Mn1401	Traill	29.6075	14.5798	10.81	2.03	0.0676	0.2333
leaflet 3	Mn1801	Traill	51.4250	14.5798	10.81	3.53	0.0049	0.0202
leaflet 7	Hendrick	Mn1401	−38.1513	15.0225	11.77	−2.54	0.0263	0.1063
leaflet 7	Hendrick	Mn1801	−60.6463	15.0225	11.77	−4.04	0.0017	0.0085
leaflet 7	Hendrick	Traill	−2.1163	15.0225	11.77	−0.14	0.8903	0.9989
leaflet 7	Mn1401	Mn1801	−22.4950	14.5798	10.81	−1.54	0.1516	0.4459
leaflet 7	Mn1401	Traill	36.0350	14.5798	10.81	2.47	0.0314	0.1186
leaflet 7	Mn1801	Traill	58.5300	14.5798	10.81	4.01	0.0021	0.0089

about the within subjects error model. In a split plot model, the within subjects errors or split plot errors are assumed to be independent and normally distributed. In a repeated measures model they are assumed to be normally distributed, but not necessarily independent.

4.6 SELECTION OF A COVARIANCE MODEL

There are many potential models for the covariance structure in a linear mixed model. Each represents an attempt to approximate the relationship between observations or errors associated with experimental units. Some of the commonly used covariance models in agricultural applications are described below.

- Unstructured model: All variances and covariances are arbitrary. This model allows for unequal variances and unequal covariances among the observations.

- Independence model: All variances are equal, and all covariances are zero. Under the normality assumption, this is equivalent to assuming that observations are independent. It represents the opposite extreme from an unstructured model.

- Compound symmetry: All variances are equal and all covariances are equal. This model represents the simplest non-trivial covariance structure.

For applications in which observations can be ordered, for example, by time or space, several additional covariance models are available. The distinction between the distance between factor levels in the list of ordered levels and the number of distance units between consecutive factor levels is important. A difference between consecutive factor levels may be more than one unit of distance. Moreover, the number of distance units between consecutive factor levels may not be the same for all pairs of consecutive levels. For example, suppose that observations are made at three times (factor levels), say at Days 0, 1, and 4. The first two factor levels differ by 1 day (distance unit), but the second and third factor levels differ by 3 days (distance units).

- First order autoregressive model: All variances are equal and the correlation between observations at adjacent factor levels is ρ, between observations two factor levels apart is ρ^2, etc. This model assumes that the factor levels are equally spaced at the same distance between consecutive levels.

- Toeplitz model: All variances are equal and the correlation between observations at adjacent factor levels is ρ_1, between observations two factor levels apart is ρ_2, etc. This model is somewhat less restrictive than an autoregressive model since; for example, ρ_2 does not have to equal ρ_1^2.

- Power model: All variances are equal and the correlation between observations at a distance of d units apart is ρ^d, where ρ is the correlation between observations at a distance of one unit.

- First order antedependence: Variances at different times are unequal, and the covariance between two observations is the product of the correlations between each adjacent time and the product of the standard deviations for the two observations. This model shares features of the unstructured and first order autoregressive models. As in the unstructured model, both the variance and the correlation between adjacent observations change with time. As in the first order autoregressive model, correlations between observations two or more time units apart are the product of the standard deviations of the observations and the correlations between observations at all times between them.

Under residual maximum likelihood estimation (REML), selection of a covariance model can be addressed through comparisons of the information criteria for various models (Section 3.3) or by formal hypothesis testing of the covariance parameters. In general, if the selected covariance structure under-models the true correlation (e.g., the independence model is used when there is non-negligible correlation), the type I error rate will be inflated. If the selected covariance structure over-models the true correlation (e.g., the unstructured model is used when a first order autoregressive model would be adequate), then the power suffers. Hence, it is important to identify the simplest covariance model that adequately accounts for the correlation structure in the data. This maximizes power without compromising control over type I error.

Information criteria such as AICC or BIC (Section 3.3) can be used to compare covariance structures provided that the fixed effects portion of the model is the same for all covariance structures under consideration. The information criteria are computed for each candidate model, and their values are compared. The candidate models should include all covariance structures deemed to be scientifically plausible in the context of the data being analyzed. This requires an understanding of the process under investigation. It is not a purely statistical question of choosing the model with the smallest value of the information criteria.

Formal comparison of two covariance structures that are nested (i.e., one can be obtained from the other by placing restrictions on some or all of the covariance parameters) can be accomplished using a likelihood ratio test based on REML (Section 2.7). The likelihood ratio test statistic formed by −2 times the natural logarithm of the ratio of the residual likelihoods from the models being compared has approximately a χ^2 distribution if one of the models is a subset of the other. The degrees of freedom for the χ^2 statistic is equal to the difference between the number of covariance parameters in the general model and the subset model. For example, to compare the unstructured model with the independence model for a 3×3 covariance matrix, the latter is a subset or special case of the former obtained by setting the covariances to zero and setting all variances equal in the unstructured model. These restrictions on the variances and covariances form the null hypothesis that is being tested. The unstructured model has six covariance parameters, and the independence model has one. Hence, the χ^2 statistic will have five degrees of freedom.

4.7 A REPEATED MEASURES EXAMPLE

In this section, an experiment comparing calcium concentrations at various soil depths is used to illustrate the analysis of repeated measures data, including selection of an appropriate covariance structure.

EXAMPLE 4.5

Root growth deep into the soil profile enables plants to better utilize soil moisture and nutrients. However, root growth in the subsoil can be inhibited by factors such as calcium deficiency and aluminum toxicity. Beyrouty et al. (2000) conducted a study of this problem on soils from the Southern Mississippi Valley. In this example, a subset of the calcium data from that study will be analyzed.

Six soil series from the loessial plains landscape that had tested for low calcium (*ca*) concentration were selected. For each soil series (*series*), five soil cores from the top 75 cm were obtained and sectioned into 15-cm depth increments.

Extractable calcium (mg/kg) was measured on each section. One observation from one section of a soil core was missing. The design was a one factor (*series*) completely randomized design with a soil core as the experimental unit. The 15-cm core sections (*depth*) represent a repeated measures factor. Both soil series and depth were assumed to be fixed factors. In repeated measures terminology, a soil core was the subject, the one factor design for soil series represents the between subjects model, and the depth portion represents the within subjects model.

Assuming that calcium concentration is approximately normally distributed, the conditional mean of the concentration, Y, is given by

$$E[Y_{ijk} \mid w_{j(i)}] = \beta_0 + S_i + w_{j(i)} + D_k + SD_{ik} \text{ for } i = 1, \ldots, 6; j = 1, \ldots, 5; k = 1, \ldots, 5$$

where β_0 is the overall mean, S_i is the ith soil series effect, D_k is the kth depth effect, SD_{ik} is the soil series $\times$ depth interaction effect, and $w_{j(i)}$ is the effect of the jth replication within the ith soil effect. Equivalently, $w_{j(i)}$ is the random error term for the between subjects model.

The model can be written in matrix form as

$$E[Y \mid u] = X\beta + Zu$$

where the random effects are multivariate normal with $u \sim MVN(0, G)$. The vector β contains the overall mean and the fixed soil series and depth effect parameters as well their interaction parameters. X and Z are the design matrices. Since u is the vector of random errors for the between subjects model, $G = \sigma_w^2 I$, where I is the identity matrix and σ_w^2 is the variance of the between subjects errors. The R matrix contains the covariance structure of the repeated measures or within subjects model. In general, R will not be a diagonal matrix.

The analysis will proceed in two steps. First, a suitable covariance structure will be selected for the within subjects model. Then the entire model will be fit using the selected covariance structure, and inference will proceed as usual.

The choice of covariance structures was narrowed down to two finalists by fitting models with various within subject covariance models. Selection of a candidate set of models was based on the small sample corrected Akaike information criterion (AICC). As an example, the GLIMMIX statements to fit the unstructured covariance model are shown in Fig. 4–21. In the *random* statement, *depth* indicates

FIG. 4–21. GLIMMIX statements to fit the model with an unstructured covariance structure for the within subjects model for Example 4.5.

```
proc glimmix data=rm;
    class series rep depth;
    model ca=series depth depth*series / ddfm=kr;
    random depth / residual type=un subject=rep(series);
run;
```

TABLE 4.2. Small sample corrected Akaike information criteria (AICC) for selected covariance structures for the within subjects model for Example 4.5. Smaller AICC values indicate better fitting models.

Covariance structure	GLIMMIX TYPE option	AICC
Independent, equal variances	–	1508.79
Compound symmetry	cs	1508.79
First order autoregressive	ar(1)	1507.70
Toeplitz, only $\rho_1 \neq 0$	toep(2)	1510.21
Independent, unequal variance	un(1)	1488.00
Heterogeneous compound symmetry	csh	1488.35
Heterogeneous AR(1)	arh(1)	1492.64
Heterogeneous Toep(2)	toeph(2)	1493.83
First order ante-dependence	ante(1)	1493.41
Unstructured	un	1496.13

the repeated measures factor and the *residual* option indicates that the statement applies to the **R** matrix that contains the within subject covariance model. The *type* option specifies the covariance structure to be fitted. For SAS users who are familiar with PROC MIXED, the *random* statement replaces the *repeated* statement, which does not exist in GLIMMIX.

The results are summarized in Table 4–2. The unstructured covariance model and the independent with equal variances models were included because they represent the extremes in complexity. As expected based on Example 4.4 and the comments following it, the independence and compound symmetry covariance structure models had the same AICC. Their heterogeneous variance versions [*un(1)* and *csh*, respectively] also had essentially the same AICC value. The independence model with unequal variances was obtained by fitting an unstructured model and forcing all covariances to be set to zero. The estimated correlation in the first order autoregressive model was 0.26. Since all other covariances are based on powers of this correlation, it is not surprising that it had an AICC similar to the Toeplitz structure with only one non-zero correlation. The AICCs for the heterogeneous versions of these two covariance structures are similar as well. The heterogeneous independence model and the heterogeneous compound symmetry have the smallest AICC values and will be examined further.

To complete the selection of a within subject covariance structure, formal hypothesis tests for non-zero covariances and unequal variances will be performed. The *covtest* statement in GLIMMIX constructs likelihood ratio based tests for covariance parameters. It has several built-in tests that are identified by keyword options. For example, the keyword *diagr* tests conditional independence in the **R** matrix; that is, it tests the null hypothesis that all covariances in the **R** matrix are zero. In addition, the user can create tests for hypotheses of specific interest in the study using the keyword *general* followed by a set of coef-

FIG. 4-22. GLIMMIX statements to fit the model with a heterogeneous compound symmetry covariance structure for the within subjects model in Example 4.5.

```
proc glimmix data=rm;
   class series rep depth;
   model ca=series depth depth*series / ddfm=kr;
   random depth / residual type=csh subject=rep(series);
   covtest 'Zero covariances' diagR / estimates;
run;
```

FIG. 4-23. GLIMMIX output containing basic model and fitting information assuming a heterogeneous compound symmetry covariance structure for the within subjects model for Example 4.5.

Model Information

Data Set	WORK.RM
Response Variable	ca
Response Distribution	Gaussian
Link Function	Identity
Variance Function	Default
Variance Matrix Blocked By	rep(series)
Estimation Technique	Restricted Maximum Likelihood
Degrees of Freedom Method	Kenward–Roger
Fixed Effects SE Adjustment	Kenward–Roger

Class Level Information

Class	Levels	Values
series	6	3 4 7 10 13 16
rep	5	1 2 3 4 5
depth	5	1 2 3 4 5

Number of Observations Read	149
Number of Observations Used	149

Dimensions

R-side Cov. Parameters	6
Columns in X	42
Columns in Z per Subject	0
Subjects (Blocks in V)	30
Max Obs per Subject	5

ficients corresponding to the covariance parameters in the entire model in the order specified in the *Covariance Parameter Estimates* section of the output.

The GLIMMIX statements to fit a heterogeneous compound symmetry covariance structure are shown in Fig. 4–22. The *covtest* statement tests the null hypothesis $H_0: \rho = 0$, which is equivalent to setting all covariances to zero. The basic model information is shown in Fig. 4–23. The 30 subjects correspond to the 30 soil cores, and

FIG. 4–24. GLIMMIX output containing the estimated covariance parameters and test for correlation equal to zero assuming a heterogeneous compound symmetry covariance structure for the within subjects model for Example 4.5.

Covariance Parameter Estimates

Cov Parm	Subject	Estimate	Standard Error
Var(1)	rep(series)	20280	5618.05
Var(2)	rep(series)	27773	8415.50
Var(3)	rep(series)	10974	3005.44
Var(4)	rep(series)	4652.36	1383.16
Var(5)	rep(series)	6701.27	1875.48
CSH	rep(series)	0.3654	.

Tests of Covariance Parameters Based on the Restricted Likelihood

					Estimates H0						
Label	DF	−2 Res Log Like	ChiSq	Pr > ChiSq	Est1	Est2	Est3	Est4	Est5	Est6	Note
Zero covariance	1	1497.45	21.85	<0.0001	20280	27773	10974	4652	6701	0	DF

DF: P-value based on a chi-square with DF degrees of freedom.

the five observations per subject are the five depth measurements. The correlation and the variances of the five depths are the six R-side covariance parameters.

Figure 4.24 shows the estimated covariance parameters and the results of the likelihood ratio test for H_0: $\rho = 0$. The variance estimates for the individual depths range from 4652 to 27,773. The estimated correlation was 0.3654. The p-value for the restricted likelihood ratio chi-squared test was less than 0.0001, providing very strong evidence that the correlation was significantly different from zero. The variances for the individual depths if the null hypothesis were true are given under *Est1* through *Est5*, and the correlation under *Est6* is zero as specified in the null hypothesis.

The missing standard error for the estimated correlation in Fig. 4–24 is an indication that there may be computational problems that could affect the results. In some situations, missing standard errors of covariance parameters may be an artifact of the scale on which the data were recorded. To investigate this possibility, the calcium concentration values were rescaled by dividing by 10, and the model was refit. The block of the **V** matrix corresponding to the first soil core for series 3, the estimated covariance parameters, and the test for zero covariances are shown in Fig. 4–25. The rescaling appears to have fixed the missing standard error problem. Comparing these results to Fig. 4–24, the estimates are similar, and differences can be attributed to the rescaling of the observations to reduce the magnitude of the data values. For the remainder of this example, the rescaled calcium values will be used.

FIG. 4-25. GLIMMIX output containing the first block of the **V** matrix, the estimated covariance parameters, and test for correlation equal to zero assuming a heterogeneous compound symmetry covariance structure for the within subjects model for Example 4.5 using the rescaled observations.

Estimated V Matrix for rep(series) 1 3

Row	Col1	Col2	Col3	Col4	Col5
1	215.43	87.9209	60.6270	36.2599	45.5977
2	87.9209	256.51	66.1558	39.5666	49.7560
3	60.6270	66.1558	121.97	27.2836	34.3098
4	36.2599	39.5666	27.2836	43.6295	20.5201
5	45.5977	49.7560	34.3098	20.5201	68.9943

Covariance Parameter Estimates

Cov Parm	Subject	Estimate	Standard Error
Var(1)	rep(series)	215.43	63.2805
Var(2)	rep(series)	256.51	70.9732
Var(3)	rep(series)	121.97	36.8759
Var(4)	rep(series)	43.6295	12.0832
Var(5)	rep(series)	68.9943	19.8563
CSH	rep(series)	0.3740	0.1026

Tests of Covariance Parameters Based on the Restricted Likelihood

Label	DF	−2 Res Log Like	ChiSq	Pr > ChiSq	Est1	Est2	Est3	Est4	Est5	Est6	Note
							Estimates H0				
Zero covariance	1	949.43	22.21	<.0001	203	278	110	46.5	67	0	DF

DF: P-value based on a chi-square with DF degrees of freedom.

Next we consider the second candidate model. The GLIMMIX statements to fit the heterogeneous independence model are shown in Fig. 4–26. The first *random* statement specifies the between subjects covariance structure of the model while the second *random* statement specifies the independent heterogeneous within subjects covariance structure. The *covtest* statement uses one of the built-in tests to test the null hypothesis

$$H_0 : \sigma^2_{D1} = \sigma^2_{D2} = \sigma^2_{D3} = \sigma^2_{D4} = \sigma^2_{D5}$$

where σ^2_{Di} indicates the within subjects model variance for the ith depth, $i = 1, \ldots, 5$.

FIG. 4-26. GLIMMIX statements to fit the heterogeneous independence model for Example 4.5 using the rescaled observations.

```
proc glimmix data=rm10 plots=studentpanel(BLUP);
    class series rep depth;
    model ca = series depth depth*series / ddfm=kr;
    random intercept / subject=rep(series) G;
    random _residual_ / type=vc group=depth V;
    covtest 'common within subj var' homogeneity / estimates;
    lsmeans depth*series / slice=series slicediff=series adjust=Tukey
        plot=meanplot(sliceby=series join);
    lsmeans depth*series / slice=depth slicediff=depth adjust=Tukey
        plot=meanplot(sliceby=depth cl);
    output out=new pred=pred stderr=sepred resid=resid student=student;
run;
```

FIG. 4-27. GLIMMIX output containing basic model and fitting information assuming a heterogeneous independence structure for the within subjects model in Example 4.5 using the rescaled observations.

Model Information

Data Set	WORK.RM10
Response Variable	ca
Response Distribution	Gaussian
Link Function	Identity
Variance Function	Default
Variance Matrix Blocked By	rep(series)
Estimation Technique	Restricted Maximum Likelihood
Degrees of Freedom Method	Kenward–Roger
Fixed Effects SE Adjustment	Kenward–Roger

Class Level Information

Class	Levels	Values
Series	6	3 4 7 10 13 16
Rep	5	1 2 3 4 5
depth	5	1 2 3 4 5

Number of Observations Read	149
Number of Observations Used	149

Dimensions

G-side Cov. Parameters	1
R-side Cov. Parameters	5
Columns in X	42
Columns in Z per Subject	1
Subjects (Blocks in V)	30
Max Obs per Subject	5

Figure 4–27 contains the basic model information. The G-side parameter is σ_w^2, and the five R-side parameters are the depth variances from the within subjects structure.

FIG. 4–28. GLIMMIX output containing the **G** matrix, the first block of the **V** matrix, the estimated covariance parameters, and test for correlation equal to zero assuming a heterogeneous independence covariance structure for the within subjects model for Example 4.5 using the rescaled observations.

Estimated G Matrix

Effect	Row	Col1
Intercept	1	34.5603

Estimated V Matrix for rep(series) 1 3

Row	Col1	Col2	Col3	Col4	Col5
1	200.00	34.5603	34.5603	34.5603	34.5603
2	34.5603	216.59	34.5603	34.5603	34.5603
3	34.5603	34.5603	138.70	34.5603	34.5603
4	34.5603	34.5603	34.5603	45.7194	34.5603
5	34.5603	34.5603	34.5603	34.5603	72.6406

Covariance Parameter Estimates

Cov Parm	Subject	Group	Estimate	Standard Error
Intercept	rep(series)		34.5603	12.7329
Residual (VC)		depth 1	165.44	49.7059
Residual (VC)		depth 2	182.03	54.7274
Residual (VC)		depth 3	104.14	33.2211
Residual (VC)		depth 4	11.1591	7.5667
Residual (VC)		depth 5	38.0803	13.2660

Tests of Covariance Parameters Based on the Restricted Likelihood

Label	DF	−2 Res Log Like	ChiSq	Pr > ChiSq	Est1	Est2	Est3	Est4	Est5	Est6	Note
common within subj var	4	956.67	29.44	<0.0001	46.4	94.5	94.5	94.5	94.5	94.5	DF

DF: P-value based on a chi-square with DF degrees of freedom.

The estimated covariance parameters, the estimated **G** matrix and the first block of the **V** matrix corresponding to the first core from soil series 3 are shown in Fig. 4–28 along with the results from the *covtest* statement. In the *Covariance Parameter Estimates* table, the first row is the estimate of σ_w^2 and the remaining rows contain the estimates of the σ_{Di}^2. The diagonal elements of the **V** matrix are the sum of the estimates of σ_w^2 and σ_{Di}^2. The off-diagonal terms represent the covariance between observations from the same soil core at different depths. The

FIG. 4–29. GLIMMIX output containing the results of the fixed effects F-tests assuming a heterogeneous independence covariance structure for the within subjects model for Example 4.5 using the rescaled observations.

Type III Tests of Fixed Effects

Effect	Num DF	Den DF	F Value	Pr > F
series	5	35.36	10.69	<0.0001
depth	4	41.81	279.92	<0.0001
series*depth	20	59.32	14.56	<0.0001

likelihood ratio test based on the REML estimates is highly significant, indicating that the within subjects variances differ by depth. *Est1* estimates σ_w^2 under $H_{0'}$ and the remaining estimates correspond to the common value of the σ_{Di}^2 under H_0.

Comparing the variance estimates in the **V** matrix block in Fig. 4–28 with the corresponding values in Fig. 4–25, the two fitted models produce roughly the same values. In addition, the average of the off-diagonal elements of the **V** matrix block is slightly larger than the corresponding common value in Fig. 4–28. Hence, the covariance structures of the models are similar, and either model could be used for the analysis. The heterogeneous independence model will be used in the remainder of this example.

Based on the results of the fixed effects F-tests in Fig. 4–29, there was a significant soil series × depth interaction. The least squares means are shown in Fig. 4–30. Note that the standard errors differ by depth (within subjects) but are the same for all soil series at the same depth (between subjects) except for depth 3 in soil series 13, where there was a missing observation in the first soil core. The larger value for the standard error for this mean reflects a smaller sample size. The tests for the simple effects by soil series and depth are presented in Fig. 4–31 with the corresponding graphs of least squares means in Fig. 4–32. The results of the comparisons among the corresponding means using Tukey's procedure are not shown.

Finally, the panel of plots of conditional studentized residuals is shown in Fig. 4–33. The quantile plot indicates that the normality assumption was not unreasonable. There appears to be an outlier that was identified as core 4 at depth 3 from soil 13. It had a studentized residual of 3.68. The effect of this observation on the results could be studied by removing it from the data set and rerunning the analyses. Given the relatively large sample size and a studentized residual that is not excessively large, we would not anticipate drastic changes in the results. ∎

4.8 ANALYSIS OF COVARIANCE

The term analysis of covariance is used in two different, but related contexts. In analysis of variance, the procedure serves to adjust the responses for uncontrolled quantitative variables before comparison of the treatment means. The uncontrolled nuisance variables are referred to as covariates or concomitant variables. For example, in an experiment to compare crop yields for different varieties, the

FIG. 4-30. GLIMMIX output containing the least squares means assuming a heterogeneous in-dependence covariance structure for the within subjects model for Example 4.5 using the rescaled observations.

series*depth Least Squares Means

Soil number	15 cm depth increment	Estimate	Standard Error	DF	t Value	Pr > |t|
3	1	1124.60	63.2651	30.25	17.78	<0.0001
3	2	677.10	65.8275	31.05	10.29	<0.0001
3	3	426.10	52.6597	28.48	8.09	<0.0001
3	4	345.90	30.2396	25.93	11.44	<0.0001
3	5	413.90	38.1124	29.78	10.86	<0.0001
4	1	947.10	63.2651	30.25	14.97	<0.0001
4	2	596.90	65.8275	31.05	9.07	<0.0001
4	3	509.70	52.6597	28.48	9.68	<0.0001
4	4	594.20	30.2396	25.93	19.65	<0.0001
4	5	781.70	38.1124	29.78	20.51	<0.0001
7	1	1797.80	63.2651	30.25	28.42	<0.0001
7	2	815.40	65.8275	31.05	12.39	<0.0001
7	3	387.30	52.6597	28.48	7.35	<0.0001
7	4	334.70	30.2396	25.93	11.07	<0.0001
7	5	328.70	38.1124	29.78	8.62	<0.0001
10	1	1065.60	63.2651	30.25	16.84	<0.0001
10	2	666.10	65.8275	31.05	10.12	<0.0001
10	3	403.50	52.6597	28.48	7.66	<0.0001
10	4	300.50	30.2396	25.93	9.94	<0.0001
10	5	309.60	38.1124	29.78	8.12	<0.0001
13	1	959.30	63.2651	30.25	15.16	<0.0001
13	2	938.90	65.8275	31.05	14.26	<0.0001
13	3	439.88	57.7672	27.91	7.61	<0.0001
13	4	292.90	30.2396	25.93	9.69	<0.0001
13	5	363.20	38.1124	29.78	9.53	<0.0001
16	1	893.90	63.2651	30.25	14.13	<0.0001
16	2	462.50	65.8275	31.05	7.03	<0.0001
16	3	207.20	52.6597	28.48	3.93	0.0005
16	4	226.20	30.2396	25.93	7.48	<0.0001
16	5	332.90	38.1124	29.78	8.73	<0.0001

stand count (number of plants per unit area) may vary from plot to plot. If there is a relationship between yield and stand count, analysis of covariance would be used to adjust the mean yields before they are compared across varieties. Roughly speaking, this is accomplished by using the yield–stand count relationship to pre-dict the mean yield at a common stand count for all varieties and then comparing the predicted mean yields at that common stand count.

FIG. 4-31. GLIMMIX output containing the results of the tests for the simple effects by soil series and by depth increment assuming a heterogeneous independence covariance structure for the within subjects model for Example 4.5 using the rescaled observations.

Tests of Effect Slices for series*depth Sliced By series				
Soil number	Num DF	Den DF	F Value	Pr > F
3	4	41.75	45.80	<0.0001
4	4	41.75	17.36	<0.0001
7	4	41.75	157.24	<0.0001
10	4	41.75	46.71	<0.0001
13	4	42.03	52.95	<0.0001
16	4	41.75	33.82	<0.0001

Tests of Effect Slices for series*depth Sliced By depth				
15 cm depth increment	Num DF	Den DF	F Value	Pr > F
1	5	30.3	28.42	<0.0001
2	5	31.08	6.41	0.0003
3	5	28.39	3.69	0.0107
4	5	25.93	17.70	<0.0001
5	5	29.79	22.33	<0.0001

Analysis of covariance is also used in studies where the objective is to compare the regression relationship between a response and a set of independent variables or predictors for several populations, each defined by a different treatment. In this context, the covariates are the predictors, and the treatment effects on the response allow the regression coefficients to differ by population. For example, it may be of interest to model soil pH as a function of extractable calcium and magnesium for soils from different texture classes and to determine whether the relationship depends on the texture. The two covariates are extractable calcium and magnesium, and the populations are defined by the soil texture classes.

Formally, an analysis of covariance model is a linear model that has at least one qualitative and one quantitative predictor. Although statistical methods textbooks often only present models with a single covariate whose coefficient does not depend on the treatments (i.e., no covariate × treatment interaction term is included), there is no limit to the number of covariates or to their functional form in the model. In addition, the design for the qualitative predictor portion of the model is not restricted to one factor (fixed) completely randomized or randomized complete block designs. Milliken and Johnson (2002) provided an extensive discussion of analysis of covariance for linear mixed models.

FIG. 4-32. GLIMMIX output displaying the least squares means for depth increment by soil series and by depth increment assuming a heterogeneous independence covariance structure for the within subjects model for Example 4.5 using the rescaled observations.

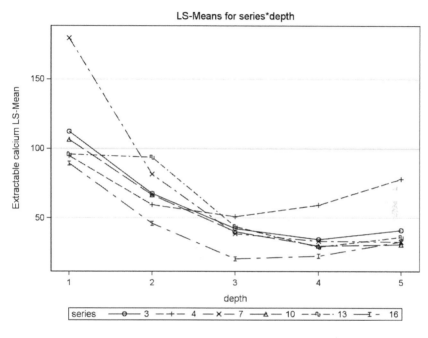

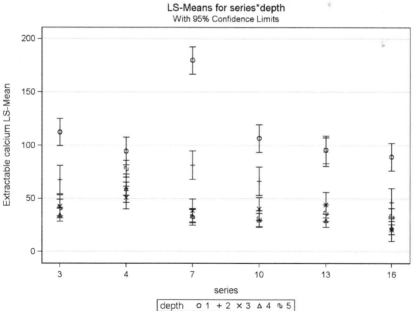

FIG. 4-33. GLIMMIX output displaying the panel of graphs of conditional studentized residuals assuming a heterogeneous independence covariance structure for the within subjects model for Example 4.5 using the rescaled observations.

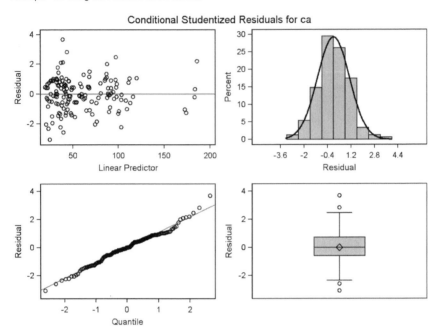

EXAMPLE 4.6

Test weight (grain weight per volume) is an important factor for grading wheat in the United States. It is influenced by kernel plumpness, which in turn, reflects the environmental conditions under which the wheat was grown. Low test weight may result in a lower sale price for the producer. An ideal variety (trial entry) is one that is high yielding and has a high test weight.

Data provided by J.T. Kelly (used with permission) from the Arkansas Wheat Variety Trials will be used to illustrate an analysis of covariance approach to address the following questions:

Is there evidence of a yield–test weight relationship?

If there is a relationship, does it differ by trial entry?

Four entries from one location of the 2006 through 2009 Variety Trials were selected for analysis. The field design in each year was a randomized complete block with four blocks. Both test weight (*testwt*) and grain yield (*yield*) were measured in the field on each plot as it was being harvested. Yield was not available for one plot in 2007. Entry was assumed to be a fixed effect. Years and blocks were treated as random effects. Years were chosen to be random instead of fixed because they represent a sample of environmental conditions under which the wheat could be grown. Yield, years, and blocks were assumed to be normally distributed.

Scatter plots of yield (bu/ac) versus test weight (lb/bu) for each entry separately indicated a generally positive linear relationship. The yield Y_{ijk} for the ith entry in the kth block within the jth year can be modeled as

$$Y_{ijk} = \beta_{0ijk} + \beta_{1ijk} W_{ijk} \text{ for } i = 1, \ldots, 4; j = 1, \ldots, 4; k = 1, \ldots, 4$$

where β_{0ijk} is the intercept, β_{1ijk} is the slope of the line, and W_{ijk} is the test weight.

The intercept and slope can be decomposed to account for possible entry, year, and block effects; that is,

$$\beta_{0ijk} = \beta_0 + E_{0i} + S_{0j} + ES_{0ij} + B_{0k(j)}$$

and

$$\beta_{1ijk} = \beta_1 + E_{1i} + S_{1j}$$

where β_0 and β_1 are the overall intercept and slope, respectively; E_{0i} and E_{1i} are the entry effects; S_{0j} and S_{1j} are the year (seasonal) effects; ES_{0ij} is the entry × year interaction effect on the intercept, and $B_{0k(j)}$ is the block within year effect on the intercept.

Note that the random block within year and entry × year interaction effects were not included in the decomposition of the slope. The variance components for test weight × block within year and test weight × entry × block within year and the residual variance would not be identifiable if they were included.

Combining the above equations, the conditional mean yield, Y, is given by

$$E[Y_{ijk} \mid S_{0j}, S_{1j}, B_{0j(k)}] = \beta_0 + E_{0i} + S_{0j} + ES_{0ij} + B_{0k(j)} + (\beta_1 + E_{1i} + S_{1j})W_{ijk}$$

In matrix form, the model can be expressed as

$$E[\mathbf{Y} \mid \mathbf{u}] = \mathbf{X\beta} + \mathbf{Zu} \text{ and } var[\mathbf{Y} \mid \mathbf{u}] = \mathbf{R}$$

where $\mathbf{u} \sim MVN(\mathbf{0}, \mathbf{G})$.

The vector β contains the 10 fixed effects parameters (overall intercept + 4 entry effects on the intercept + overall slope + 4 entry effects on the slope), and $\mathbf{u}$ contains the 40 random effects (4 year effects on the intercept + 16 block within year effects on the intercept + 16 year × entry effects on the intercept + 4 year effects on the slope). $\mathbf{X}$ is the 63 × 10 fixed effects design matrix, and $\mathbf{Z}$ is the 63 × 40 random effects design matrix. The four parameters in the covariance matrix $\mathbf{G}$ are the variances of the year effect on the intercept, the blocks within year effect on the intercept, the year × entry interaction effect on the intercept, and the year effect on the slope.

The GLIMMIX statements for the initial fit of the model are given in Fig. 4–34, and the basic model information is shown in Fig. 4–35. Since test weight is a quantitative predictor, it does not appear in the *class* statement. Defining year as the

FIG. 4-34. GLIMMIX statements to fit the initial analysis of covariance model for Example 4.6.

```
proc glimmix data=wheat;
    class block year entry;
    model yield = entry testwt entry*testwt / ddfm=kr;
    random intercept block entry testwt / subject=year;
    covtest 'random slope coeff = 0'   . . . 0 ;
run;
```

FIG. 4-35. GLIMMIX output containing basic model and fitting information for the initial fit for Example 4.6.

Model Information

Data Set	WORK.WHEAT
Response Variable	yield
Response Distribution	Gaussian
Link Function	Identity
Variance Function	Default
Variance Matrix Blocked By	year
Estimation Technique	Restricted Maximum Likelihood
Degrees of Freedom Method	Kenward–Roger
Fixed Effects SE Adjustment	Kenward–Roger

Class Level Information

Class	Levels	Values
block	4	1 2 3 4
year	4	2006 2007 2008 2009
entry	4	214 247 275 458

Number of Observations Read	64
Number of Observations Used	63

Dimensions

G-side Cov. Parameters	4
R-side Cov. Parameters	1
Columns in X	10
Columns in Z per Subject	10
Subjects (Blocks in V)	4
Max Obs per Subject	16

subject produces four blocks (or sets) of columns in **Z**, one per year, with 10 columns per block (year effect on the intercept + 4 block within year effects on the intercept + 4 year × entry effects on the intercept + year effect on the slope) and 16 (or 15 for 2007) observations per subject. The *covtest* statement in Fig. 4–34 tests the hypothesis that the variance of the year effect on the slope is zero. The statement contains one coefficient for each variance component in the model with the order of the coefficients following their order in the *Covariance Parameter Estimates* section of the output.

FIG. 4–36. GLIMMIX output containing the estimated variance components and results of the tests of significance for the initial fit for Example 4.6.

Covariance Parameter Estimates

Cov Parm	Subject	Estimate	Standard Error
Intercept	year	2.44E-17	.
block	year	33.8422	17.6088
entry	year	69.7343	39.6933
testwt	year	0.01042	0.01752
Residual		23.7337	6.1012

Tests of Covariance Parameters Based on the Restricted Likelihood

Label	DF	−2 Res Log Like	ChiSq	Pr > ChiSq	Note
random slope coeff = 0	1	410.33	0.01	0.9332	–

–: Standard test with unadjusted p-values.

Figure 4–36 gives the estimated variance components and the result of the variance component test. Note that the variance of the year effect is essentially zero (2.44×10^{-17}) for the intercept and very small for the slope relative to the remaining variance components. The p-value for the test of a zero variance for the year effect on the slope is 0.9332, from which we conclude that there is no evidence of unequal slopes by year and the year × test weight term can be dropped from the model for the slope. At this point, we suspend judgment about the variance of the year effect on the intercept. In random coefficient models, negative and zero variance component estimates can be artifacts of negligible unequal slope random effects, and the issue often disappears once the random slope term is removed from the model.

To refit the model without year × test weight, the test weight term in the *random* statement in Fig. 4–34 was deleted. The estimated covariance parameters and the tests for the fixed effects are shown in Fig. 4–37. The variance of the year effect on the intercept is no longer zero, and all estimated variances have standard error estimates. The test of the fixed effects indicates that there is a significant entry effect on the slope ($p = 0.0472$); that is, there are significant differences among the slopes. The *entry* effect tests the hypothesis that the intercepts do not depend on entry; that is, it tests the hypothesis

$$H_0: E_{01} = E_{02} = E_{03} = E_{04} = 0$$

This is not a test of equal entry means in any agronomically meaningful sense. Literally, it is a test of equal entry means given that the test weight is zero, which clearly has no valid interpretation in the context of the example. Tests of effects on

FIG. 4-37. GLIMMIX output containing the estimated variance components and tests for the fixed effects for the model with year × test weight removed in Example 4.6.

Covariance Parameter Estimates

Cov Parm	Subject	Estimate	Standard Error
Intercept	year	29.7755	50.1573
block	year	33.8561	17.6253
entry	year	70.5109	39.9461
Residual		23.7620	6.1081

Type III Tests of Fixed Effects

Effect	Num DF	Den DF	F Value	Pr > F
entry	3	38.31	3.10	0.0378
testwt	1	42.57	23.67	<0.0001
testwt*entry	3	38.49	2.90	0.0472

FIG. 4-38. GLIMMIX statements to fit the model with year × test weight removed for Example 4.6.

```
proc glimmix data=wheat plots=studentpanel;
    class block year entry;
    model yield = entry entry*testwt / ddfm=kr noint solution;
    random intercept block entry / subject=year;
    lsmeans entry / diff at testwt=55.6313;    /* entry effect @ overall mean          */
    lsmeans entry / diff at testwt=58.0;       /* entry effect @ dockage cutoff        */
    lsmeans entry / diff at testwt=57.0828;    /* entry effect @ 275 - 458 intersection */
    lsmeans entry / diff bylevel e;            /* individual entry means               */
run;
```

the intercepts are often not interpreted if any of the effects on the covariate coefficients are significant; that is, the goal of the analysis is to determine the significant effects on the slope coefficients.

The GLIMMIX statements to fit the model and to obtain estimates of the slope and intercept coefficients are shown in Fig. 4–38. In the *model* statement the *testwt* term has been removed, and the *noint* and *solution* options have been added. The *noint* option instructs GLIMMIX to exclude the overall intercept β_0 from the fitted model. The absence of the *testwt* term indicates that β_1 has also been removed from the model. These two changes eliminate the decomposition of the fixed effects portion of the intercept and slope and change the interpretation of E_{0i} and E_{1i}. The *solution* option requests that the estimated regression coefficients be included in the output.

FIG. 4-39. GLIMMIX output containing the results of the tests for the fixed effects and estimated coefficients for the model with year × test weight removed in Example 4.6.

Solutions for Fixed Effects

Effect	Entry number	Estimate	Standard Error	DF	t Value	Pr > |t|
entry	214	−49.4696	70.9333	42.57	−0.70	0.4893
entry	247	−172.69	73.7641	43.7	−2.34	0.0239
entry	275	−167.10	45.8675	46.08	−3.64	0.0007
entry	458	24.2872	59.1510	38.85	0.41	0.6836
testwt*entry	214	2.1278	1.2418	42.6	1.71	0.0939
testwt*entry	247	4.3552	1.2928	43.65	3.37	0.0016
testwt*entry	275	4.4256	0.8289	45.43	5.34	<0.0001
testwt*entry	458	1.0728	1.0885	39.08	0.99	0.3304

Type III Tests of Fixed Effects

Effect	Num DF	Den DF	F Value	Pr > F
Entry	4	40.9	4.89	0.0026
testwt*entry	4	41.06	10.06	<0.0001

Figure 4–39 contains the results of the tests of the fixed effects and the estimated regression coefficients. For example, the fixed effects component of the predicted yield for entry 247 is given by $-172.69 + 4.36W_{2jk}$. The p-value associated with the slope estimate indicates that there is sufficient evidence to indicate a statistically significant non-zero linear coefficient. Note that the tests for fixed effects are not the same as those in Fig. 4–37. For example, the hypothesis tested by *entry* × *testwt* is

$$H_0: E_{11} = E_{12} = E_{13} = E_{14}$$

whereas the corresponding test in Fig. 4–37 is for the hypothesis in which these effects equal zero.

The least squares means are the points on the estimated lines at a given test weight. If there were no significant entry effect on the slope, the differences among the entries would be the same for all test weights within the range found in the data, and differences would be determined by comparing the least squares means at the mean test weight. These means are the traditional analysis of covariance adjusted means.

In general, when there is a significant treatment × covariate interaction, there are two types of analyses that can be used to describe the differences. First, the coefficients of the covariate can be compared among treatment levels. In GLIMMIX this can be accomplished using *contrast* or *estimate* statements. Second, the least

FIG. 4-40. Graph of the fitted lines from the fit of the model with year × test weight removed in Example 4.6. The lines are identified by the entry number.

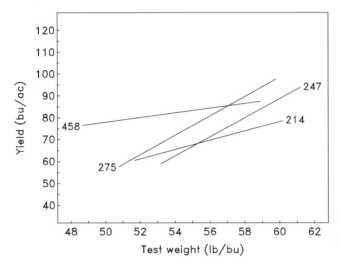

squares means (or predicted values of the response) can be compared at a series of covariate values that are of interest from a subject matter point of view. The latter analysis will be illustrated in this example.

The fitted lines have been plotted on the same graph in Fig. 4–40 using the estimated coefficients in Fig. 4–39. From the figure it is clear that differences in predicted values (or least squares means) will depend on test weight. Test weights at which comparison of entries may be of interest would include the overall mean test weight (55.63), the test weight at which producers could receive a lower price for their wheat (58.0), and the test weight (approximately 57.08) at which entries 275 and 458 have the same predicted yield. This latter test weight was obtained by setting the equations for the fitted lines equal and solving for test weight. Finally, it may be of interest to compare predicted yields for entries at their individual mean test weights rather than at a common test weight since it is not reasonable to assume a priori that all entries would have the same mean test weight.

The *lsmeans* statements to obtain the above comparisons are shown in Fig. 4–38. The *bylevel* option in the last *lsmeans* statement specifies that the mean test weight for each individual entry should be used in the calculation for that entry. The *e* option requests that these mean test weights be displayed in the output.

The least squares means and their differences at the overall mean test weight are presented in Fig. 4–41. The comparisons at the dockage test weight are shown in Fig. 4–42. In Fig. 4–43, the least squares means for entries 275 and 458 agree to two decimal places. Finally, in Fig. 4–44, the first table shows the entry test weight means used in the least squares means calculations. Although there are very few statistically significant differences in these comparisons, they do serve to illustrate the method by which specific objectives can be addressed. ■

FIG. 4-41. GLIMMIX output comparing least squares means at the overall mean test weight for Example 4.6.

entry Least Squares Means

Entry number	Margins	testwt	Estimate	Standard Error	DF	t Value	Pr > |t|
214	Balanced	55.63	68.9019	5.6018	8.487	12.30	<0.0001
247	Balanced	55.63	69.5999	5.6026	8.545	12.42	<0.0001
275	Balanced	55.63	79.1009	5.3838	7.64	14.69	<0.0001
458	Balanced	55.63	83.9692	5.6174	8.545	14.95	<0.0001

Differences of entry Least Squares Means

Entry number	Entry number	Margins	testwt	Estimate	Standard Error	DF	t Value	Pr > |t|
214	247	Balanced	55.63	-0.6980	6.6194	8.524	-0.11	0.9185
214	275	Balanced	55.63	-10.1990	6.4438	8.175	-1.58	0.1513
214	458	Balanced	55.63	-15.0673	6.6396	8.882	-2.27	0.0498
247	275	Balanced	55.63	-9.5010	6.4251	8.186	-1.48	0.1766
247	458	Balanced	55.63	-14.3693	6.6801	9.001	-2.15	0.0599
275	458	Balanced	55.63	-4.8682	6.4329	8.125	-0.76	0.4706

The importance of Example 4.6 is that it illustrates that questions of practical interest can be addressed even when the treatment effects on the covariate are not the same across all treatments. Moreover, in general, these conclusions will be different than those that would have been obtained from an incorrectly simplified model obtained by removing, or initially not considering, the treatment and year effects on the covariate.

4.9 BEST LINEAR UNBIASED PREDICTION

The conditional mean of Y given the random effects **u** represents an average for a particular set of values of the random variables u_j. For example, the conditional mean yield for an entry in Example 4.6 depends on the effects of a sample of years and blocks within years. Even though these effects are random, it is sometimes of interest to predict their values for specific levels of the populations of years and blocks from which the samples were drawn. This is accomplished by using their best linear unbiased predictors (BLUPs). Since these are random variables, this is different than estimating the value of a parameter such as an entry effect on the slope coefficients in the example.

FIG. 4-42. GLIMMIX output comparing least squares means at the dockage test weight for Example 4.6.

entry Least Squares Means

Entry number	Margins	testwt	Estimate	Standard Error	DF	t Value	Pr > |t|
214	Balanced	58.00	73.9420	5.5093	8.113	13.42	<0.0001
247	Balanced	58.00	79.9162	5.5382	8.269	14.43	<0.0001
275	Balanced	58.00	89.5839	5.9193	10.32	15.13	<0.0001
458	Balanced	58.00	86.5104	6.8332	13.57	12.66	<0.0001

Differences of entry Least Squares Means

Entry number	Entry number	Margins	testwt	Estimate	Standard Error	DF	t Value	Pr > |t|
214	247	Balanced	58.00	−5.9741	6.4821	8.154	−0.92	0.3832
214	275	Balanced	58.00	−15.6419	6.7296	9.407	−2.32	0.0440
214	458	Balanced	58.00	−12.5683	7.5636	11.19	−1.66	0.1243
247	275	Balanced	58.00	−9.6677	6.8182	9.624	−1.42	0.1878
247	458	Balanced	58.00	−6.5942	7.5212	11.18	−0.88	0.3991
275	458	Balanced	58.00	3.0735	7.8321	12.83	0.39	0.7012

FIG. 4-43. GLIMMIX output comparing least squares means at the test weight where entries 275 and 458 have equal least squares means for Example 4.6.

entry Least Squares Means

Entry number	Margins	testwt	Estimate	Standard Error	DF	t Value	Pr > |t|
214	Balanced	57.08	71.9904	5.3570	7.516	13.44	<0.0001
247	Balanced	57.08	75.9215	5.3596	7.527	14.17	<0.0001
275	Balanced	57.08	85.5247	5.6374	8.851	15.17	<0.0001
458	Balanced	57.08	85.5264	6.2653	11.15	13.65	<0.0001

Differences of entry Least Squares Means

Entry number	Entry number	Margins	testwt	Estimate	Standard Error	DF	t Value	Pr > |t|
214	247	Balanced	57.08	−3.9311	6.1891	7.37	−0.64	0.5445
214	275	Balanced	57.08	−13.5343	6.4245	8.271	−2.11	0.0671
214	458	Balanced	57.08	−13.5360	6.9834	9.689	−1.94	0.0822
247	275	Balanced	57.08	−9.6032	6.4322	8.283	−1.49	0.1725
247	458	Balanced	57.08	−9.6048	6.9750	9.665	−1.38	0.1996
275	458	Balanced	57.08	−0.00166	7.1640	10.51	−0.00	0.9998

FIG. 4-44. GLIMMIX output comparing least squares means at the mean test weights of individual entries for Example 4.6.

Coefficients for entry Least Squares Means

Effect	Entry number	Row1	Row2	Row3	Row4
entry	214	1			
entry	247		1		
entry	275			1	
entry	458				1
testwt*entry	214	56.956			
testwt*entry	247		56.906		
testwt*entry	275			54.956	
testwt*entry	458				53.706

entry Least Squares Means

Entry number	Margins	Estimate	Standard Error	DF	t Value	Pr > |t|
214	WORK.WHEAT	71.7211	5.3547	7.507	13.39	<0.0001
247	WORK.WHEAT	75.1526	5.3547	7.507	14.03	<0.0001
275	WORK.WHEAT	76.1134	5.3547	7.507	14.21	<0.0001
458	WORK.WHEAT	81.9040	5.3894	7.647	15.20	<0.0001

Differences of entry Least Squares Means

Entry number	Entry number	Margins	Estimate	Standard Error	DF	t Value	Pr > |t|
214	247	WORK.WHEAT	-3.4315	6.1827	7.353	-0.56	0.5954
214	275	WORK.WHEAT	-4.3923	6.1827	7.353	-0.71	0.4994
214	458	WORK.WHEAT	-10.1828	6.2127	7.448	-1.64	0.1426
247	275	WORK.WHEAT	-0.9608	6.1827	7.353	-0.16	0.8807
247	458	WORK.WHEAT	-6.7513	6.2127	7.448	-1.09	0.3111
275	458	WORK.WHEAT	-5.7905	6.2127	7.448	-0.93	0.3805

EXAMPLE 4.7

This is a continuation of Example 4.6. In that example, the results indicated that the slopes of the lines differed by entry but were not affected by the random effects and the intercepts differed by entry subject to additional significant year and entry × year random effects. Based on these conclusions, the regression equations for each entry represent a set of parallel lines with intercepts varying by year; that is, the relationship (slope) between yield and test weight would be the same each year, but the overall yield level and potentially the range of test weights would vary by year. Differences among the slopes of different entries could be determined using *estimate* or *contrast* statements as described in Example 4.6.

An added objective to those in Example 4.6 would be to examine the overall yield level represented by the random intercepts. Averaging over blocks within years, the year specific regression model would be described by the random intercept model whose linear predictor is given by

$$(\beta_0 + E_{0i} + S_{0j} + ES_{0ij}) + (\beta_1 + E_{1i})W_{ijk}$$

The year specific intercepts are combinations of fixed and random effects and can be predicted by their best linear unbiased predictors (BLUPs).

To make the programming easier, the version of the model used in the final fit in Example 4.6 where the fixed effects portion of the coefficients were not decomposed will be used here. The SAS statements to obtain the predicted intercepts are shown in Fig. 4–45. Two *random* statements are used since we are not interested in BLUPs at the block within year level. The *solution* option on the first *random* statement produces predictions of the year and entry × year effect on the intercepts.

The BLUPs for the regression coefficients are produced by *estimate* statements. For a particular regression coefficient, the fixed effects and their sets of coefficients are listed first, separated by a vertical bar from the random effects and their coefficients. The *subject* option determines the specific subject (*year* in this example) to be used in the calculation.

For example, the first *estimate* statement in Fig. 4–45 produces the BLUP for the intercept associated with the first entry (214) in the first year (2006). Because the *noint* option was used on the *model* statement, the fixed effect *entry* portion of the statement combined with the *subject* option estimates the sum $\beta_0 + E_{01}$. The random effect portion specified by *intercept* and *entry* combined with the *subject* option predicts S_{01} and ES_{011}, respectively.

The predicted values of the random year and entry × year effects on the intercepts are shown in Fig. 4–46, and the BLUPs are presented in Fig. 4–47. For example, for entry 214 in 2006, the predicted year effect is 4.2066, and the predicted entry × year effect is 7.9290. Note that the predicted intercept is obtained by the adding the entry effect from Fig. 4–39 and the random effects from Fig. 4–46 (i.e., −49.4696 + 4.2066 + 7.9290 = −37.3340), which agrees with the value in Fig. 4–47.

The predicted intercepts can be compared across years within an entry or across entries within a year. As an example, the *estimate* statements in Fig. 4–45 will provide the predicted differences and a test of the hypothesis of no difference for the first entry (214) across all years and all entries within the first year (2006). For the within entry comparisons of years, there is no fixed effect term in the *estimate* statement because it is the same for all years and cancels out in the difference. The *subject* option indicates the years being compared. For the entry comparisons within the same year, the random year effect will cancel and is not included in the *random* effects portion of the *estimate* statement. The predicted differences are shown in Fig. 4–48. The only significant differences at the 0.05 level between

FIG. 4-45. GLIMMIX statements to fit the final model from Example 4.6 and to obtain best linear unbiased predictions of the regression coefficients in Example 4.7.

```
proc glimmix data=wheat;
  class block year entry;
  model yield = entry testwt / ddfm=kr noint solution;
  random intercept entry / subject=year solution;
  random block / subject=year;
  estimate 'Intercept entry 1 year 1' entry 1 0 0 0 | intercept 1 entry 1 0 0 0 / subject 1 0 0 0;
  estimate 'Intercept entry 1 year 2' entry 1 0 0 0 | intercept 1 entry 1 0 0 0 / subject 0 1 0 0;
  estimate 'Intercept entry 1 year 3' entry 1 0 0 0 | intercept 1 entry 1 0 0 0 / subject 0 0 1 0;
  estimate 'Intercept entry 1 year 4' entry 1 0 0 0 | intercept 1 entry 1 0 0 0 / subject 0 0 0 1;

  estimate 'Intercept entry 2 year 1' entry 0 1 0 0 | intercept 1 entry 0 1 0 0 / subject 1 0 0 0;
  estimate 'Intercept entry 2 year 2' entry 0 1 0 0 | intercept 1 entry 0 1 0 0 / subject 0 1 0 0;
  estimate 'Intercept entry 2 year 3' entry 0 1 0 0 | intercept 1 entry 0 1 0 0 / subject 0 0 1 0;
  estimate 'Intercept entry 2 year 4' entry 0 1 0 0 | intercept 1 entry 0 1 0 0 / subject 0 0 0 1;

  estimate 'Intercept entry 3 year 1' entry 0 0 1 0 | intercept 1 entry 0 0 1 0 / subject 1 0 0 0;
  estimate 'Intercept entry 3 year 2' entry 0 0 1 0 | intercept 1 entry 0 0 1 0 / subject 0 1 0 0;
  estimate 'Intercept entry 3 year 3' entry 0 0 1 0 | intercept 1 entry 0 0 1 0 / subject 0 0 1 0;
  estimate 'Intercept entry 3 year 4' entry 0 0 1 0 | intercept 1 entry 0 0 1 0 / subject 0 0 0 1;

  estimate 'Intercept entry 4 year 1' entry 0 0 0 1 | intercept 1 entry 0 0 0 1 / subject 1 0 0 0;
  estimate 'Intercept entry 4 year 2' entry 0 0 0 1 | intercept 1 entry 0 0 0 1 / subject 0 1 0 0;
  estimate 'Intercept entry 4 year 3' entry 0 0 0 1 | intercept 1 entry 0 0 0 1 / subject 0 0 1 0;
  estimate 'Intercept entry 4 year 4' entry 0 0 0 1 | intercept 1 entry 0 0 0 1 / subject 0 0 0 1;

  estimate 'Intercept difference entry 1, yr 1 vs yr 2' | intercept 1 entry 1 -1 0 0 / subject 1 -1 0 0;
  estimate 'Intercept difference entry 1, yr 1 vs yr 3' | intercept 1 entry 1 0 -1 0 / subject 1 0 -1 0;
  estimate 'Intercept difference entry 1, yr 1 vs yr 4' | intercept 1 entry 1 0 0 -1 / subject 1 0 0 -1;
  estimate 'Intercept difference entry 1, yr 2 vs yr 3' | intercept 1 entry 0 1 -1 0 / subject 0 1 -1 0;
  estimate 'Intercept difference entry 1, yr 2 vs yr 4' | intercept 1 entry 0 1 0 -1 / subject 0 1 0 -1;
  estimate 'Intercept difference entry 1, yr 3 vs yr 4' | intercept 1 entry 0 0 1 -1 / subject 0 0 1 -1;

  estimate 'Intercept difference entry 1 vs entry 2, yr 1' entry 1 -1 0 0 | entry 1 -1 0 0 / subject 1 0 0 0;
  estimate 'Intercept difference entry 1 vs entry 3, yr 1' entry 1 0 -1 0 | entry 1 0 -1 0 / subject 1 0 0 0;
  estimate 'Intercept difference entry 1 vs entry 4, yr 1' entry 1 0 0 -1 | entry 1 0 0 -1 / subject 1 0 0 0;
  estimate 'Intercept difference entry 2 vs entry 3, yr 1' entry 0 1 -1 0 | entry 0 1 -1 0 / subject 1 0 0 0;
  estimate 'Intercept difference entry 2 vs entry 4, yr 1' entry 0 1 0 -1 | entry 0 1 0 -1 / subject 1 0 0 0;
  estimate 'Intercept difference entry 3 vs entry 4, yr 1' entry 0 0 1 -1 | entry 0 0 1 -1 / subject 1 0 0 0;

  contrast 'Intercept difference entry 1, yr 1 vs yr 2' | intercept 1 entry 1 -1 0 0 / subject 1 -1 0 0;
  contrast 'Intercept difference entry 1, yr 1 vs yr 3' | intercept 1 entry 1 0 -1 0 / subject 1 0 -1 0;
  contrast 'Intercept difference entry 1, yr 1 vs yr 4' | intercept 1 entry 1 0 0 -1 / subject 1 0 0 -1;
  contrast 'Intercept difference entry 1, yr 2 vs yr 3' | intercept 1 entry 0 1 -1 0 / subject 0 1 -1 0;
  contrast 'Intercept difference entry 1, yr 2 vs yr 4' | intercept 1 entry 0 1 0 -1 / subject 0 1 0 -1;
  contrast 'Intercept difference entry 1, yr 3 vs yr 4' | intercept 1 entry 0 0 1 -1 / subject 0 0 1 -1;

  contrast 'Intercept difference entry 1 vs entry 2, yr 1' entry 1 -1 0 0 | entry 1 -1 0 0 / subject 1 0 0 0;
  contrast 'Intercept difference entry 1 vs entry 3, yr 1' entry 1 0 -1 0 | entry 1 0 -1 0 / subject 1 0 0 0;
  contrast 'Intercept difference entry 1 vs entry 4, yr 1' entry 1 0 0 -1 | entry 1 0 0 -1 / subject 1 0 0 0;
  contrast 'Intercept difference entry 2 vs entry 3, yr 1' entry 0 1 -1 0 | entry 0 1 -1 0 / subject 1 0 0 0;
  contrast 'Intercept difference entry 2 vs entry 4, yr 1' entry 0 1 0 -1 | entry 0 1 0 -1 / subject 1 0 0 0;
  contrast 'Intercept difference entry 3 vs entry 4, yr 1' entry 0 0 1 -1 | entry 0 0 1 -1 / subject 1 0 0 0;
run;
```

FIG. 4-46. GLIMMIX output containing the predicted values for the random year and entry × year effects on the intercepts for Example 4.7.

Solution for Random Effects

Effect	Entry number	Subject	Estimate	Std Err Pred	DF	t Value	Pr > |t|
Intercept		year 2006	4.2066	5.9497	1.498	0.71	0.5732
Entry	214	year 2006	7.9290	6.2057	9.687	1.28	0.2311
Entry	247	year 2006	9.2836	6.1997	9.667	1.50	0.1662
Entry	275	year 2006	−10.3720	6.2235	9.784	−1.67	0.1272
Entry	458	year 2006	3.1210	6.2238	9.737	0.50	0.6272
Intercept		year 2007	1.3709	6.0501	1.459	0.23	0.8484
Entry	214	year 2007	3.2669	6.5256	10.57	0.50	0.6269
Entry	247	year 2007	−4.5957	6.4747	10.54	−0.71	0.4932
Entry	275	year 2007	0.06028	6.3797	10.36	0.01	0.9926
Entry	458	year 2007	4.5149	7.0361	11.23	0.64	0.5340
Intercept		year 2008	−0.6351	5.9910	1.481	−0.11	0.9281
Entry	214	year 2008	−4.8024	6.3421	10.1	−0.76	0.4662
Entry	247	year 2008	−0.3083	6.4292	10.41	−0.05	0.9627
Entry	275	year 2008	1.4405	6.4725	10.93	0.22	0.8280
Entry	458	year 2008	2.1664	6.2661	9.959	0.35	0.7367
Intercept		year 2009	−4.9425	6.1569	1.415	−0.80	0.5347
Entry	214	year 2009	−6.3935	7.1385	11.94	−0.90	0.3881
Entry	247	year 2009	−4.3796	6.9952	11.99	−0.63	0.5430
Entry	275	year 2009	8.8712	6.6152	11.6	1.34	0.2056
Entry	458	year 2009	−9.8023	6.9419	11.27	−1.41	0.1849

years for the first entry are years one and three. In the first year comparisons, the intercept for entry four (458) is different from entries two and three (247 and 275).

The *contrast* statements in Fig. 4–45 also provide tests for the differences in predicted intercepts. The format used to specify a particular difference is the same as that in the corresponding *estimate* statement. Figure 4–49 contains the results of the single degree of freedom contrast F-tests. Since these F-tests have only one numerator degree of freedom, their p-values are identical to the p-values for the corresponding t-tests. In fact, for single degree of freedom contrasts, $t^2 = F$. ∎

FIG. 4-47. GLIMMIX output containing the best linear unbiased predictors (BLUPs) for the intercepts for Example 4.8.

Estimates

Label	Estimate	Standard Error	DF	t Value	Pr > |t|
Intercept entry 1 year 1	−37.3340	71.4097	42.68	−0.52	0.6038
Intercept entry 1 year 2	−44.8319	73.2908	42.36	−0.61	0.5440
Intercept entry 1 year 3	−54.9071	72.5772	42.47	−0.76	0.4535
Intercept entry 1 year 4	−60.8056	66.0770	43.53	−0.92	0.3625
Intercept entry 2 year 1	−159.20	73.2853	43.84	−2.17	0.0353
Intercept entry 2 year 2	−175.91	75.9246	43.54	−2.32	0.0253
Intercept entry 2 year 3	−173.63	75.9269	43.56	−2.29	0.0271
Intercept entry 2 year 4	−182.01	69.5557	44.23	−2.62	0.0121
Intercept entry 3 year 1	−173.27	46.5709	45.81	−3.72	0.0005
Intercept entry 3 year 2	−165.67	44.9695	45.64	−3.68	0.0006
Intercept entry 3 year 3	−166.30	48.3559	45.87	−3.44	0.0013
Intercept entry 3 year 4	−163.17	42.9890	45.57	−3.80	0.0004
Intercept entry 4 year 1	31.6149	58.3606	39.46	0.54	0.5911
Intercept entry 4 year 2	30.1730	63.6858	38	0.47	0.6384
Intercept entry 4 year 3	25.8185	59.4101	39.17	0.43	0.6663
Intercept entry 4 year 4	9.5425	54.7086	40.48	0.17	0.8624

FIG. 4-48. GLIMMIX output for differences between predicted intercepts across years for the first entry (214) and across entries for the first year (2006) in Example 4.7 using *estimate* statements.

Estimates

Label	Estimate	Standard Error	DF	t Value	Pr > |t|
Intercept difference entry 1, yr 1 vs yr 2	−6.3814	10.3480	11.74	−0.62	0.5492
Intercept difference entry 1, yr 1 vs yr 3	29.3856	10.0335	10.32	2.93	0.0146
Intercept difference entry 1, yr 1 vs yr 4	10.5483	12.0595	15.63	0.87	0.3950
Intercept difference entry 1, yr 2 vs yr 3	−0.9013	10.2257	12.53	−0.09	0.9312
Intercept difference entry 1, yr 2 vs yr 4	−8.2199	13.4677	27.8	−0.61	0.5466
Intercept difference entry 1, yr 3 vs yr 4	−15.0920	12.3080	17.1	−1.23	0.2367
Intercept difference entry 1 vs entry 2, yr 1	121.86	103.78	36.11	1.17	0.2480
Intercept difference entry 1 vs entry 3, yr 1	135.93	79.0627	40.24	1.72	0.0932
Intercept difference entry 1 vs entry 4, yr 1	−68.9488	88.6444	34.3	−0.78	0.4420
Intercept difference entry 2 vs entry 3, yr 1	14.0693	86.1496	42.38	0.16	0.8710
Intercept difference entry 2 vs entry 4, yr 1	−190.81	87.1930	36.56	−2.19	0.0351
Intercept difference entry 3 vs entry 4, yr 1	−204.88	71.4187	41.85	−2.87	0.0064

FIG. 4-49. GLIMMIX output for the single degree of freedom contrast tests for differences between predicted intercepts across years for the first entry (214) and across entries for the first year (2006) in Example 4.7.

Contrasts				
Label	Num DF	Den DF	F Value	Pr > F
Intercept difference entry 1, yr 1 vs yr 2	1	11.74	0.38	0.5492
Intercept difference entry 1, yr 1 vs yr 3	1	10.32	8.58	0.0146
Intercept difference entry 1, yr 1 vs yr 4	1	15.63	0.77	0.3950
Intercept difference entry 1, yr 2 vs yr 3	1	12.53	0.01	0.9312
Intercept difference entry 1, yr 2 vs yr 4	1	27.8	0.37	0.5466
Intercept difference entry 1, yr 3 vs yr 4	1	17.1	1.50	0.2367
Intercept difference entry 1 vs entry 2, yr 1	1	36.11	1.38	0.2480
Intercept difference entry 1 vs entry 3, yr 1	1	40.24	2.96	0.0932
Intercept difference entry 1 vs entry 4, yr 1	1	34.3	0.60	0.4420
Intercept difference entry 2 vs entry 3, yr 1	1	42.38	0.03	0.8710
Intercept difference entry 2 vs entry 4, yr 1	1	36.56	4.79	0.0351
Intercept difference entry 3 vs entry 4, yr 1	1	41.85	8.23	0.0064

REFERENCES CITED

Beyrouty, C.A., J.K. Keino, E.E. Gbur, and M.G. Hanson. 2000. Phytotoxic concentrations of subsoil aluminum as influenced by soils and landscape position. Soil Sci. 165:135–143. doi:10.1097/00010694-200002000-00004

Giesbrecht, F.G., and J.C. Burns. 1985. Two stage analysis based on a mixed model: Large sample asymptotic theory and small sample simulation results. Biometrics 41:477–486. doi:10.2307/2530872

Henderson, C.R. 1984. Applications of linear models in animal breeding. University of Guelph, Guelph, Ontario, Canada.

Kackar, R.N., and D.A. Harville. 1984. Approximations for standard errors of estimators of fixed and random effects in mixed linear models. J. Am. Stat. Assoc. 79:853–862. doi:10.2307/2288715

Kenward, M.G., and J.H. Roger. 1997. Small sample inference for fixed effects from restricted maximum likelihood. Biometrics 53:983–997. doi:10.2307/2533558

McLean, R.A., W.L. Sanders, and W.W. Stroup. 1991. A unified approach to mixed linear models. Am. Stat. 45:54–64. doi:10.2307/2685241

Milliken, G.A., and D.E. Johnson. 2002. Analysis of messy data. Volume III: Analysis of covariance. Chapman and Hall/CRC Press, Boca Raton, FL.

Robinson, G.K. 1991. That BLUP is a good thing: The estimation of random effects. Stat. Sci. 6:15–51. doi:10.1214/ss/1177011926

Satterthwaite, F.E. 1946. An approximate distribution of estimates of variance components. Biometrics 2:110–114. doi:10.2307/3002019

Searle, S.R. 1971. Linear models. John Wiley and Sons, New York.

Seversike, T.M., L.C. Purcell, E.E. Gbur, P. Chen, and R. Scott. 2009. Radiation interception and yield response to increased leaflet number in early-maturing soybean genotypes. Crop Sci. 49:281–289. doi:10.2135/cropsci2007.08.0472

Stroup, W.W., and R.C. Littell. 2002. Impact of variance component estimates on fixed effect inference in unbalanced linear mixed models. p. 32–48. In Proceedings of the Conference on Applied Statistics in Agriculture. Dep. of Statistics, Kansas State University, Manhattan, KS.

GENERALIZED LINEAR MIXED MODELS

5.1 INTRODUCTION

Generalized linear mixed models combine the generalized linear models introduced in Chapter 3 with the linear mixed models described in Chapter 4. As an extension of generalized linear models, they incorporate random effects into the linear predictor. As a mixed model, they contain at least one fixed effect and at least one random effect.

More specifically, let Y be the response variable whose conditional distribution given the random effects belongs to the exponential family or can be written as a quasi-likelihood. Let $x_1,..., x_p$ be a set of p explanatory variables describing the fixed effects and let $u_1,..., u_q$ be a set of q random effects. The linear predictor of the model for the jth observation given the random effects is expressed as

$$\eta_{ij} = g\left(E\left[Y_j \mid u_1,..., u_q\right]\right) = \beta_0 + \sum_{i=1}^{p} \beta_i x_{ij} + \sum_{k=1}^{q} z_{kj} u_k, j = 1,...,n$$

where β_0 is the overall mean, β_i is the ith fixed effect coefficient, x_{ij} is the ith fixed effect explanatory variable on the jth observation, z_{kj} is the binary indicator variable for the effect of the kth random effect, u_k, on the jth observation, and $g(\cdot)$ is the link function relating the conditional mean of the response to the predictors.

Comparing the above specification to Section 4.1, the response is no longer required to be normally distributed, and the relationship between the conditional mean of the response and the linear predictor is now on the link scale. However, the data remain on the original or data scale (see Section 3.1). This is an important distinction between these models and those where the data are transformed before analysis.

In matrix form, the linear predictor can be written as

$$\eta = g(E[Y \mid u]) = X\beta + Zu$$

where Y is the $n \times 1$ vector of responses, X is the $n \times (p + 1)$ fixed effects design matrix, β is the $(p + 1) \times 1$ vector of fixed effects coefficients, Z is the $n \times q$ design matrix

doi:10.2134/2012.generalized-linear-mixed-models.c5

Analysis of Generalized Linear Mixed Models in the Agricultural and Natural Resources Sciences
Edward E. Gbur, Walter W. Stroup, Kevin S. McCarter, Susan Durham, Linda J. Young, Mary Christman, Mark West, and Matthew Kramer

for the random effects and $\mathbf{u}$ is the $q \times 1$ vector of random effects. The conditional variance can be written as

$$\text{var}[\mathbf{Y} \mid \mathbf{u}] = \mathbf{R} = \phi \mathbf{V}^{1/2} \mathbf{P} \mathbf{V}^{1/2},$$

where $\mathbf{P}$ is a working correlation matrix, $\mathbf{V}^{1/2}$ is a diagonal matrix with the square root of the variance function on the diagonal, and ϕ is a scale parameter. When $\mathbf{P}$ is the identity matrix, then $\mathbf{R}$ is an $n \times n$ covariance matrix. The distribution of $\mathbf{u}$ is multivariate normal with mean $\mathbf{0}$ and covariance matrix $\mathbf{G}$; i.e., $\mathbf{u} \sim \text{MVN}(\mathbf{0}, \mathbf{G})$.

5.2 ESTIMATION AND INFERENCE IN GENERALIZED LINEAR MIXED MODELS

Inference in generalized linear mixed models involves estimation and testing of the unknown parameters in β, $\mathbf{R}$, and $\mathbf{G}$ as well as prediction of the random effects $\mathbf{u}$. As in generalized linear models, all inference is performed on the link scale (or model scale) and not on the data scale of the response variable. Reporting results on the original scale requires converting model scale estimates to the data scale using the inverse of the link function. Approximate standard errors are obtained using the delta method described in Section 3.2.

Estimation in generalized linear mixed models is based on maximum likelihood (Section 2.4). The two basic computational approaches to obtain solutions to the likelihood equations are pseudo-likelihood and integral approximation of the log-likelihood using either Laplace or Gauss–Hermite quadrature methods. The pseudo-likelihood (PL) approach is applicable to a broader range of models than the Laplace and quadrature methods. In addition, a pseudo-likelihood adaptation of the Kenward–Roger correction can be used to adjust the standard error estimates and test statistics, and although it is ad hoc, it appears to be accurate as long as the linear approximation used by PL is accurate. However, pseudo-likelihood suffers from two drawbacks. First, it produces biased covariance parameter estimates when the number of observations per subject is small, as is the case in many agricultural applications, and is particularly prone to biased estimates when the power is small. This problem appears to be exacerbated for two parameter distributions; i.e., for the negative binomial, beta and gamma distributions. Second, since PL uses a pseudo-likelihood instead of a true likelihood, likelihood ratio and fit statistics such as AICC or BIC have no clear meaning. Thus, competing models cannot be compared using likelihood ratio tests or information criteria. Since the Laplace and quadrature approaches use the actual likelihood, they do not suffer from this drawback.

The Laplace and quadrature approaches are applicable for a smaller range of models than pseudo-likelihood. For example, neither of these methods can be used for models that include R-side random effects. In certain cases, these effects can be rewritten as G-side effects, allowing Laplace and quadrature to be used. Both methods require conditionally independent observations. In addition, the quadrature method requires that the random effects be processed by subject; i.e., the *subject* option on the *random* statement in GLIMMIX must be used. If there is more than one random effect, they must form a containment hierarchy. The Laplace method is more flexible in this regard.

For both the Laplace and quadrature methods, as the number of random effects increases, the procedures become more computationally intensive and eventually will become prohibitive. Of the two methods, quadrature becomes computationally prohibitive more quickly as the model complexity and/or size increases. Since the marginal likelihood is approximated numerically in these methods, some aspects of the marginal distribution of the response are not available. For example, the estimated marginal covariance matrix of the response, $\mathbf{V}$, cannot be computed. The advantage of the Laplace and quadrature approaches is that the estimates tend to have better asymptotic behavior and less small sample bias than estimates obtained by pseudo-likelihood. Although the Kenward–Roger correction is not available with these methods, empirical or sandwich estimators of the covariance parameters provide an alternative. However, these alternatives introduce a different bias issue that has not been fully researched for sample sizes typically found in many agricultural experiments. This issue is discussed in more detail in Section 5.9.

5.3 CONDITIONAL AND MARGINAL MODELS

As with linear mixed models, generalized linear mixed models (GLMM) can be formulated as conditional or marginal models. This distinction was introduced in Section 2.10 and discussed in more detail for the linear mixed model in Section 4.3. Specific applications involving conditional and marginal linear mixed models for split-plot and repeated measures experiments appeared in Sections 4.4 and 4.7, respectively. Each of these applications has a generalized linear mixed model analog. In addition, various conditional and marginal generalized linear mixed models are used to account for or adjust for over-dispersion, an issue unique to models for non-normally distributed data. Over-dispersion is discussed in detail in Sections 5.5 through 5.7.

Recall that for linear mixed models, the conditional model has

- a linear predictor that includes random model effects, i.e.,
  $\eta = E[\mathbf{Y} \mid \mathbf{u}] = \mathbf{X}\boldsymbol{\beta} + \mathbf{Z}\mathbf{u},$
- random effects $\mathbf{u}$ that are assumed to be multivariate normal, i.e.,
  $\mathbf{u} \sim MVN(\mathbf{0}, \mathbf{G}),$
- data that are assumed to be conditionally multivariate normal, i.e.,
  $\mathbf{Y} \mid \mathbf{u} \sim MVN(\mathbf{X}\boldsymbol{\beta} + \mathbf{Z}\mathbf{u}, \mathbf{R}).$

The marginal linear mixed model has

- a linear predictor that includes fixed effects only, i.e., $\eta = E[\mathbf{Y}] = \mathbf{X}\boldsymbol{\beta},$
- data that are assumed to be multivariate normal, i.e., $\mathbf{Y} \sim MVN(\mathbf{X}\boldsymbol{\beta}, \mathbf{V}),$
  where $\mathbf{V} = \mathbf{Z}\mathbf{G}\mathbf{Z}' + \mathbf{R}.$

In the conditional generalized linear mixed model, the linear predictor and random model effects are defined as they were for the linear mixed model. The only difference is that the conditional distribution of the data Y on the random effects is assumed to belong to the family of distributions described in Chapter 2 for GLMMs and the linear predictor is on the link scale rather than on the data scale. Note that a conditional generalized linear mixed model must be defined on a probability distribution. If the model is defined on a quasi-likelihood, it is by definition not a conditional model.

The marginal generalized linear mixed model shares the same linear predictor as the marginal linear mixed model but the distribution and variance assumptions differ. The "distribution" applies exclusively to the data Y because there are no random effects on which to condition. Distribution appears in quotes because all marginal GLMMs are defined on quasi-likelihoods, not on true probability distributions. To account for the additional variance–covariance elements that the conditional model includes via the distribution of the random effects **u**, the marginal generalized linear mixed model uses a working correlation matrix. The working correlation matrix borrows its structure from the linear mixed model's **ZGZ′** and **R** matrices, but it is not a true correlation matrix. The result is that that the variance of the observations **Y** is modeled by

$$\mathrm{Var}\big(\mathbf{Y}\big) = \phi \mathbf{V}_\mu^{1/2} \mathbf{P} \mathbf{V}_\mu^{1/2}$$

where $\mathbf{V}_\mu^{1/2} = \mathrm{diag}[\sqrt{\mathbf{V}(\mu)}]$, $\mathbf{V}(\mu)$ is the variance of the assumed distribution, and P is the working correlation matrix.

Because working correlation matrices imply quasi-likelihood, estimation is not based on a true likelihood function. As a result, marginal GLMMs can only be estimated using pseudo-likelihood methods, whereas conditional models may use pseudo-likelihood or integral approximation. The theoretical details of quasi-likelihood estimation are not discussed here, but when practical considerations relevant to examples discussed in this chapter arise, they will be noted.

Understanding the distinction between conditional and marginal GLMMs and the issues that arise is best accomplished by revisiting the linear mixed model conditional and marginal examples in Examples 4.1 and 4.2 but working through them with a non-normal response variable.

EXAMPLE 5.1

This example is a continuation of the 10-block, two treatment randomized complete block design from Example 2.10 and discussed for normal theory linear mixed models in Examples 4.1 and 4.2. The data were given in Table 4.1. In this example the response variable is defined by $Y = F/N$, where N is the number of plants and F is the number of damaged plants. Y is assumed to have a binomial distribution. For this example the treatments will represent two varieties of a crop.

Recall that in developing the model from the ANOVA sources of variation in Example 2.10, the linear predictor was an additive function of all but the last line of the ANOVA table. Following the same reasoning used in Examples 4.1 and 4.2 in this example leads to the conditional model where

- the linear predictor is $\eta_{ij} = \beta_0 + T_i + b_j$, where β_0 is the intercept, T_i is the variety (treatment) effect, and b_j is the block effect,
- the random effect is $b_j \sim N(0, \sigma_B^2)$ and the b_j are independent,
- the conditional distribution of $Y_{ij} \mid b_j \sim$ Binomial(100, π_{ij}) and the observations Y_{ij} are independent,
- the link function is the logit; i.e., $\eta_{ij} = \log[\pi_{ij}/(1 - \pi_{ij})]$.

This is a conditional model because the distribution of the data is specified in terms of a conditional distribution given the random effects that appear in the linear predictor.

From the discussion following Example 2.10, the role of the last line in the ANOVA table must be understood differently for one parameter members of the exponential family than it is for normal theory models. For the normal theory models in Examples 4.1 and 4.2, the last line is understood as "residual." Mathematically it is identical to the block × variety interaction, which explains why there is no interaction term in the linear predictor. The last line in the ANOVA table cannot simultaneously estimate residual variance and block × variety interaction and the former must take priority. For one-parameter members of the exponential family, there is no residual variance component to estimate; the block × variety term can (and often should) appear in the linear predictor. Thus, a competing form of the linear predictor is

$$\eta_{ij} = \beta_0 + T_i + b_j + \text{Tb}_{ij}$$

where Tb_{ij} is the block × variety interaction and $\text{Tb}_{ij} \sim N(0, \sigma_{TB}^2)$ and the Tb_{ij} are assumed to be independent.

The first step in fitting the model involves deciding whether the block × treatment interaction should appear in the linear predictor. This can be done in one of two ways; either

- fit the model using the linear predictor without Tb_{ij} and use the Pearson chi-square goodness of fit statistic to check the model, or
- fit the model including Tb_{ij} in the linear predictor and use *covtest* to test H_0: $\sigma_{TB}^2 = 0$.

The GLIMMIX statements for each approach are shown in Fig. 5–1. Notice that both runs use the quadrature method because the appropriateness of including Tb_{ij} in the model must be assessed using the actual log-likelihood, not the pseudo-likelihood. Either quadrature or Laplace methods may be used but the GLIMMIX default pseudo-likelihood method cannot. Quadrature is more accurate, so it is used here. The Laplace method should be used in cases where quadrature either

FIG. 5-1. GLIMMIX statements to fit the conditional model without and with the block × variety interaction for Example 5.1.

```
title2 'CONDITIONAL MODEL WITHOUT BLOCK x VARIETY FOR EXAMPLE 5.1';
proc glimmix data=CvsM method=quadrature;
    class variety block;
    model F/N = variety;
    random intercept / subject=block;
    lsmeans variety / ilink;
run;

title2 'CONDITIONAL MODEL WITH BLOCK x VARIETY FOR EXAMPLE 5.1';
proc glimmix data=CvsM method=quadrature;
    class variety block;
    model F/N = variety;
    random intercept variety / subject=block;
    lsmeans variety / ilink cl;
    covtest 'Block x Variety = 0' . 0;
run;
```

cannot be used or is computationally prohibitive. For the binomial distribution, the response should be expressed as a ratio of the variables representing the number of damaged plants (F) divided by the total number of plants observed in the plot (N) on the *model* statement. When the response uses the syntax F/N, the binomial with the logit link is assumed but both the *link* and *dist* options can be specified if desired.

Figures 5–2 and 5–3 show the results for the Pearson chi-square conditional goodness of fit of the simpler linear predictor $\eta_{ij} = \beta_0 + T_i + b_j$ and the test of H_0: $\sigma_{TB}^2 = 0$, respectively. For the former, the Pearson chi-square is 157.59, with 20 degrees of freedom, for a ratio to its degrees of freedom of 7.88. If the model without Tb_{ij} fits well, this ratio should be approximately 1. The value 7.88 is very large compared to 1, and the p-value of a formal test would be less than 0.0001. For the second approach, the likelihood ratio chi-square is 131.24 with 1 d.f. and $p < 0.0001$; again, very strong evidence to reject H_0: $\sigma_{TB}^2 = 0$. From either approach, $\eta_{ij} = \beta_0 + T_i + b_j + Tb_{ij}$ is the appropriate linear predictor for the conditional model.

Figure 5–4 shows variance component estimates and the F-test for the fixed variety effect hypothesis H_0: $T_1 = T_2 = 0$, and the estimates on the link (logit) and data (probability) scale for variety 0 and variety 1. These results are obtained from fitting the model with the interaction included in the model. The block and block × variety variance component estimates are $\hat{\sigma}_B^2 = 1.20$ and $\hat{\sigma}_{TB}^2 = 1.10$, respectively. The block variance component is a measure of variation among block-average logits. The estimated logit is an estimate of the log odds (Example 2.1). The variety effect, $T_1 - T_2$, is the difference between the log odds, which is the log of the odds-ratio. The block × variety variance component is therefore a measure of variability of log-odds-ratios (and hence, of variety effects) among blocks. The type 3 F-test gives the approximate

FIG. 5-2. GLIMMIX output containing the Pearson chi-square goodness of fit test of the conditional model without the block × variety interaction for Example 5.1.

Fit Statistics for Conditional Distribution	
-2 log L(F | r. effects)	242.33
Pearson Chi-Square	157.59
Pearson Chi-Square / DF	7.88

FIG. 5-3. GLIMMIX output containing the test for the block × variety interaction in the conditional model for Example 5.1.

Tests of Covariance Parameters Based on the Likelihood					
Label	DF	-2 Log Like	ChiSq	Pr > ChiSq	Note
Block × Variety = 0	1	287.80	131.24	<0.0001	MI

MI: P-value based on a mixture of chi-squares.

FIG. 5-4. GLIMMIX output containing the covariance parameter estimates, test for the fixed effect, and least squares means in the conditional model with block × variety interaction for Example 5.1.

Covariance Parameter Estimates			
Cov Parm	Subject	Estimate	Standard Error
Intercept	Block	1.2044	0.9462
Variety	Block	1.0967	0.5432

Type III Tests of Fixed Effects				
Effect	Num DF	Den DF	F Value	Pr > F
Variety	1	9	6.29	0.0334

Variety Least Squares Means											
Variety	Estimate	Standard Error	DF	t Value	Pr > |t|	Alpha	Lower	Upper	Mean	Standard Error Mean	Lower Mean	Upper Mean
0	1.2839	0.4915	9	2.61	0.0282	0.05	0.1720	2.3958	0.7831	0.08348	0.5429	0.9165
1	2.5580	0.5126	9	4.99	0.0007	0.05	1.3985	3.7176	0.9281	0.03420	0.8019	0.9763

F value for testing H_0: $T_1 = T_2 = 0$. Since $F = 6.29$ and $p = 0.0334$, there is evidence at the 0.05 level of a statistically significant difference between the two varieties with regard to the probability of a damaged plant. The estimated logit for varieties 0 and 1 are

1.28 and 2.56, with standard errors 0.49 and 0.51, respectively. Applying the inverse link, the estimated probabilities for varieties 0 and 1 are 0.78 and 0.93, with standard errors 0.083 and 0.034, respectively. The confidence intervals for the probabilities are obtained by taking confidence limits for the logits and applying the inverse link to them to obtain an asymmetric confidence interval. The resulting 95% confidence intervals are [0.54, 0.92] and [0.80, 0.98] for varieties 0 and 1, respectively.

Note that if the block $\times$ variety interaction had been excluded from the model as it would have been for the normal theory version of this model, the F value for testing the variety effect would have been $F = 51.33$ with $p < 0.0001$. This illustrates an important point about the impact of omitting essential random effects in GLMMs. Omitting these effects tends to result in inflated (often severely inflated) test statistics. In this example, working through the ANOVA sources of variation to model the process shown in Example 2.10, it is clear that the unit of randomization for variety is block $\times$ variety, and hence, variety effects must be assessed relative to background random variation among block $\times$ variety units. This happens naturally in the normal theory linear models in Examples 4.1 and 4.2 because the residual variance is the measure of variation among block $\times$ variety units. Because one parameter exponential family distributions, specifically the binomial and Poisson, do not have a distinct scale parameter, variety effects will be assessed relative to the variance function of the distribution only when block $\times$ variety effects are included explicitly in the linear predictor as random effects. The result is somewhat analogous to using pseudo-replicates instead of true replicates in a conventional F-test; the usual result is inflated test statistics and excessive type I error rate. This is why mastering the ANOVA-to-model process demonstrated in Exercise 2.10 is crucial if one is to work effectively with generalized linear mixed models. This theme recurs in various forms, notably for over-dispersion and with repeated measures models. Over-dispersion is discussed in Section 5.5 and repeated measures GLMMs in Section 5. 8.

In Example 4.1, the model for the normal theory randomized block design was reformulated with a compound symmetry covariance structure. For the compound symmetry form of the conditional model in this example:

- the linear predictor is $\eta_{ij} = \beta_0 + T_i + Tb_{ij}$; i.e., the block term has been removed,

- the bivariate distribution of the random effects $[Tb_{1j}, Tb_{2j}]'$ is

$$MVN\left(0, \sigma_{CS}^2 \begin{bmatrix} 1 & \rho \\ \rho & 1 \end{bmatrix}\right),$$

- the conditional distribution of $Y_{ij} \mid Tb_{ij} \sim \text{Binomial}(100, \pi_{ij})$ and the observations Y_{ij} are independent,

- the link function is the logit; i.e., $\eta_{ij} = \log[\pi_{ij}/(1 - \pi_{ij})]$.

This is still a conditional model because the distribution of the data is conditional on the random block $\times$ variety effects. This model merely reparameterizes the block and block $\times$ variety effects into their compound symmetry form.

FIG. 5–5. GLIMMIX statements to fit the compound symmetry formulation of the conditional model in Example 5.1.

```
proc glimmix data=CvsM method=quadrature;
  class variety block;
  model F/N = variety;
  random variety / type=cs subject=block;
  lsmeans variety / ilink;
run;
```

FIG. 5–6. GLIMMIX output containing the covariance parameter estimates for the compound symmetry formulation of the conditional model for Example 5.1.

Covariance Parameter Estimates

Cov Parm	Subject	Estimate	Standard Error
Variance	Block	1.0966	0.5431
CS	Block	1.2030	0.9453

The GLIMMIX statements to fit the compound symmetry form are shown in Fig. 5–5, and the covariance parameter estimates are given in Fig. 5–6. The covariance parameter estimates are nearly identical to the results in Fig. 5–4 aside from relabeling. The fixed effect tests and least squares means results are not shown but were unchanged. ∎

Next we consider the marginal model. The marginal model has no random effects in the linear predictor and embeds all variance information in the covariance structure of the response variable. For Example 5.1, in the marginal model:

- the linear predictor is $\eta_{ij} = \mu + T_{i'}$
- there is no random component since the effects associated with the blocks are modeled as part of the covariance structure,
- the response Y_{ij} has a marginal quasi-likelihood whose form derives from the likelihood for the binomial distribution with parameter π_{ij}. However, the π_{ij} no longer have the same meaning as they did in the conditional model. The variance is modified to include a working correlation whose form is borrowed from the analogous normal distribution covariance structure, in this case compound symmetry (Section 4.3). Specifically,

$$\text{pseudo-variance}\begin{bmatrix} Y_{1j} \\ Y_{2j} \end{bmatrix} = \begin{bmatrix} \sqrt{\pi_{1j}(1-\pi_{1j})} & 0 \\ 0 & \sqrt{\pi_{2j}(1-\pi_{2j})} \end{bmatrix} \phi \begin{bmatrix} 1 & \rho \\ \rho & 1 \end{bmatrix} \begin{bmatrix} \sqrt{\pi_{1j}(1-\pi_{1j})} & 0 \\ 0 & \sqrt{\pi_{2j}(1-\pi_{2j})} \end{bmatrix}$$

This is called a pseudo-variance because a true binomial random variable cannot have this variance. For this reason, this "distribution" is actually a quasi-likelihood.

It has the form of a binomial distribution, but because of its correlation structure, it is not a true probability distribution. Note that the block covariance is embedded in the correlation structure. Superficially, this model looks like the compound symmetry form of the binomial conditional GLMM shown above. The difference is that in the conditional model, the random block × variety effect appears explicitly in the linear predictor. Here there is no random effect in the linear predictor, and the compound symmetry structure is embedded in the pseudo-variance.

The marginal generalized linear mixed model often is referred to as a generalized estimating equation (GEE) model. Strictly speaking, GEE refers to generalized linear models with no random effects in the linear predicator and all of the variance–covariance structures associated with the random factors embedded in the working correlation structure. GEEs became very popular when generalized linear mixed model computing software and computing technology in general was less developed. These models are still deeply entrenched in certain disciplines.

Technically, the GEE model fails a primary requirement of a statistical model. It should describe a plausible probability mechanism by which the observations arise. Because the quasi-likelihood is not a true probability distribution, data could never arise from the process implicit in a GEE. Nonetheless, GEEs are useful if the conditional generalized linear mixed models are too complex to be computationally tractable or if the objectives of the study are best addressed by the marginal mean rather than the conditional mean.

EXAMPLE 5.2

In this example we will fit the marginal GEE for the data used in Example 5.1. The data were given in Table 4–1.

The GLIMMIX statements to fit the marginal model are shown in Fig. 5–7. As in Fig. 5–1, the binomial response should be expressed on the *model* statement as a ratio of the variables representing the number of damaged plants and the total number of plants. As before, it is not necessary to include the *link* and *dist* options on the *model* statement for the binomial. The *type* and *subject* options on the *random* statement specify a compound symmetry covariance structure for each block. The *random* statement with the *residual* option modifies the **R** matrix by defining the form of the working correlation matrix. For non-normal data, whenever *residual* appears in the *random* statement, the model is a marginal model and the "distribu-

FIG. 5–7. GLIMMIX statements to fit the marginal (GEE) model with compound symmetry working covariance structure in Example 5.2.

```
proc glimmix data=CvsM;
  class variety block;
  model F/N = variety;
  random variety / type=cs subject=block residual ;
  lsmeans variety / ilink;
run;
```

FIG. 5–8. GLIMMIX output containing the covariance parameter estimates, test for the fixed effect, and least squares means in the marginal model with a working compound symmetry covariance structure in Example 5.2.

Covariance Parameter Estimates

Cov Parm	Subject	Estimate	Standard Error
CS	Block	4.9925	8.8440
Residual		21.0654	9.9303

Type III Tests of Fixed Effects

Effect	Num DF	Den DF	F Value	Pr > F
Variety	1	9	2.12	0.1791

Variety Least Squares Means

Variety	Estimate	Standard Error	DF	t Value	Pr > |t|	Mean	Standard Error Mean
0	0.9494	0.3599	9	2.64	0.0270	0.7210	0.07240
1	1.7036	0.4472	9	3.81	0.0042	0.8460	0.05827

tion" named in the *model* statement is a quasi-likelihood that borrows the form of the assumed distribution.

Figure 5–8 shows results analogous to those shown in Fig. 5–4 for the conditional model. The *Covariance Parameter Estimates* section shows the estimates of $\hat{\rho}\hat{\phi} = 4.99$ and $\hat{\phi} = 21.07$ for the working covariance and scale parameters, respectively. While these have analogous meanings to the block and block × variety variance components in the conditional model, here they are working covariance components, not actual variance or covariance parameters. As such, readers are strongly cautioned against attaching too literal an interpretation to them; they account for variability, but do not have any interpretation per se.

For the marginal model, the test for the variety effect is $F = 2.12$ with $p = 0.1791$ compared to the conditional model values of $F = 6.29$ with $p = 0.0334$. For the marginal model, the estimated logits are 0.95 and 1.70 for varieties 0 and 1, respectively. Their inverse-linked, data scale "probability" estimates are 0.72 and 0.85, respectively. Compare these with the conditional data scale estimated probabilities of 0.78 and 0.93. Why is there such a large discrepancy? More importantly, are these differences happenstance for this particular data set, or are they typical of systematic, predictable, and repeatable differences between the conditional and marginal models? ∎

The short answer to the questions posed in Example 5.2 is that the differences between the conditional and marginal results are not happenstance. It is easy to show that for estimated probabilities greater than 0.50, the conditional estimates will always exceed the marginal estimates and vice versa for estimated probabilities less than 0.50; that is, the conditional estimates will always be less than the marginal estimates. In addition, it can be demonstrated via simulation that the power for tests of treatment differences using the conditional model will always exceed the power of similarly defined tests using the marginal model except when the probability for each treatment is 0.50. The differences become more pronounced as the probabilities approach zero or one.

The reason that this happens lies in the probability structure of generalized linear mixed models. Recall that there are two processes generating the observations, namely, the design process (blocks) and the treatment structure (varieties). The block process follows a normal distribution. The observations on each variety, conditional on the plot in which they are observed, are binomial. However, we cannot directly observe either of these processes in isolation. We can see only the end result of both processes. The resulting counts, Y_{ij}, that are actually observed do not have a binomial distribution.

In probability distribution terms, the joint distribution of the observations, Y, the random block effect, b, and the block $\times$ variety interaction effect, Tb, is the product of the joint distribution of the random block and block $\times$ variety interaction effects and the conditional distribution of the response given the block and block $\times$ variety interaction effects. Expressing this in terms of probability distributions,

$$f(Y, b, \text{Tb}) = f(b, \text{Tb})\, f(Y \mid b, \text{Tb})$$

Only Y is directly observable. Its distribution follows from averaging out the block and block $\times$ variety interaction effects in the joint distribution. The resulting distribution is called the marginal distribution of Y. In terms of probability distributions, we have

$$f(Y) = \int_b \int_{\text{Tb}} f(Y, b, \text{Tb})$$

where each integral represents the averaging process over the distribution of that random effect.

For non-normally distributed linear models, marginal distributions, while difficult to deal with mathematically, are easy to conceptualize and to visualize using simulation. The next example demonstrates the relationship between the distributions of the estimated probabilities for the conditional and marginal models using specific values for π_{ij} in the context of Examples 5.1 and 5.2.

EXAMPLE 5.3

For illustrative purposes assume that the true values of π generating the underlying binomial models are 0.75 for variety 0 and 0.90 for variety 1. Figure 5–9 shows the marginal distribution of the sample proportion p for each variety using the assumed values of π. Both distributions are strongly left skewed.

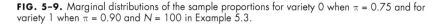

FIG. 5-9. Marginal distributions of the sample proportions for variety 0 when $\pi = 0.75$ and for variety 1 when $\pi = 0.90$ and $N = 100$ in Example 5.3.

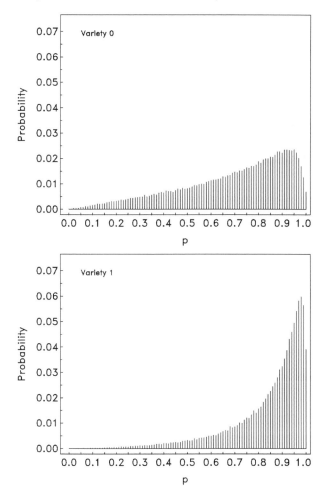

For left skewed distributions, the mean is less than the median. Assuming variances of 1.2 and 1.1 for block and block $\times$ variety, respectively (the estimates from the conditional model), for the marginal distributions shown in Fig. 5–9, the mean and median of variety 0 are 0.69 and 0.75, respectively, and for variety 1, are 0.83 and 0.90, respectively. The median of the marginal distribution equals the true probability π whereas the mean of the marginal distribution is substantially smaller.

This is where the discrepancy between the conditional GLMM (Example 5.1) and the marginal GEE model (Example 5.2) arises. The conditional GLMM is focused on estimating π for the "typical block" in the population, a "typical" block being defined as one that is neither below nor above average. This is effectively an estimate at $b = Tb = 0$. However, there is more nuance than that—one is in effect eliminating the block effects to obtain an estimate whose inference space applies to the entire population of blocks. See McLean et al. (1991) for a discussion of inference spaces in mixed models. Because $b = Tb = 0$ are the mid-points of the block effect distributions, the conditional GLMM broad inference space estimate recovers the parameter π as defined in the model statement. On the other hand, in the marginal model there is no explicit separation of the binomial distribution (which is conditional on the random effects) and the random effect distributions. Since the GEE cannot extract the binomial parameter, the best it can do is estimate the marginal mean. Since π is the primary parameter of interest, estimating the mean of the marginal distribution in this case will underestimate π.

The marginal distribution in binomial examples will be symmetric if and only if $\pi = 0.50$. Since the mean and median are equal in symmetric distributions, the mean of the marginal distribution will be π. If $\pi < 0.50$, then the marginal distribution will be right-skewed. If $\pi > 0.50$, then it will be left-skewed. In neither case will the mean and median be equal. In addition, skewness increases as π approaches zero or one. Skewness also increases as the variance of the random effects increases. Hence, the marginal mean will either over or underestimate π to a degree that depends on π and the variance components. ∎

The most important point in the above examples is that conditional and marginal models for non-normal generalized linear mixed models, unlike linear mixed models for normally distributed data, do not yield identical estimates.

The estimated probabilities from the conditional model are variously called "random effects" estimates (Molenberghs and Verbeke, 2006), "mixed model" estimates (Hardin and Hilbe, 2003) or conditional model estimates. Estimated probabilities from the marginal model are called marginal estimates or alternatively, "population averaged" (PA) estimates. Estimates from the conditional model can be understood as the estimated probability of a damaged plant one would expect for a typical or average member of the population (blocks in the examples). The marginal estimates can be understood as the mean number of damaged plants per hundred averaged over the population of blocks, assuming that the sample of blocks accurately represents the population. It is important to understand that both the conditional model and marginal model estimates lead to inferences that apply to the entire target population from which the data were drawn—in mixed model terminology, broad inference. If you want to answer the question, "How does the average block perform?" then use the conditional mixed model estimate. The marginal or PA estimate addresses the question, "How much plant damage occurs over the entire population of blocks?"

Two additional issues are worth examining before leaving these examples. These concern the normal approximation to the binomial and variance stabilizing transformations.

EXAMPLE 5.4

In traditional statistical methods courses, students are taught that when N is sufficiently large, the sample proportion (F/N in this example) can be assumed to have an approximate normal distribution. Figure 5–10 shows the GLIMMIX statements to implement the normal approximation. Note that the block effect appears in the *random* statement. The block effect is a legitimate part of the linear predictor, but block × variety cannot appear in the linear predictor because it is confounded with the residual under the approximate normality assumption.

Using the data in Table 4.1, the results are shown in Fig. 5–11. These results are identical to what would be obtained doing an analysis of variance on the proportions F/N and computing sample mean proportion for each variety. The F value for the test of equal varieties is $F = 2.42$ with a p-value of 0.1545. The estimates of the variety means, which would be interpreted as the estimates of the probabilities π_0 and π_1, are 0.72 and 0.85, respectively. These estimates are identical to the estimates obtained using the marginal GEE and the F value for testing variety effect is approximately the same. The only difference is that the standard errors using the normal approximation are equal, a result of the normal theory linear mixed model assumption of equal variances, an assumption we know must be violated by definition when the data are binomial. In this sense, the normal approximation is simply the marginal GEE done badly. ∎

FIG. 5–10. GLIMMIX statements to fit the linear mixed model for the normal approximation to the binomial in Example 5.4.

```
proc glimmix data=CvsM;
   class variety block;
   prop=F/N;
   model prop = variety;
   random intercept / subject=block;
   lsmeans variety;
run;
```

FIG. 5–11. GLIMMIX output containing the test for the fixed effect and least squares means in the normal approximation to the binomial in Example 5.4.

Type III Tests of Fixed Effects

Effect	Num DF	Den DF	F Value	Pr > F
Variety	1	9	2.42	0.1545

Variety Least Squares Means

Variety	Estimate	Standard Error	DF	t Value	Pr > |t|
0	0.7210	0.06358	9	11.34	<0.0001
1	0.8460	0.06358	9	13.31	<0.0001

Prior to the use generalized linear mixed models, the arcsine–square root transformation was the standard fix for the unequal variance issue in the normal approximation. How does the transformation perform in the context of these examples?

EXAMPLE 5.5

Figure 5–12 shows the GLIMMIX statements for the normal theory linear mixed model using the arcsine–square root transformation. The *ods output* statement captures the least squares means and related information on the transformation scale in a new data file named *lsm*. An additional data step is used to implement an inverse transformation and the delta method to provide data scale estimates and their estimated standard errors.

For the data in Table 4.1, the results are shown in Figure 5–13. The F value is 4.18 with a *p*-value of 0.0714, midway between the results for the conditional and marginal models. Similarly, the back-transformed estimates on the data scale for varieties 0 and 1 are 0.75 and 0.89, respectively, midway between the conditional and marginal estimates. ∎

The two previous examples underline the shortcomings of transformations for mixed models with non-normal data. Both the conditional and marginal models have well-defined targets of inference that can be connected with applications where each is clearly appropriate. On the other hand, results obtained via the transformation do not relate to the parameters of interest. They clearly do not estimate π or the marginal mean. Indeed, it is not clear what they estimate. Transformations do not resolve any of the issues associated with GLMM or GEE

FIG. 5–12. GLIMMIX statements to fit the linear mixed model to the binomial using the arcsine-square root transformation in Example 5.5.

```
proc glimmix data=CvsM;
   class variety block;
   transfprop=arsin(sqrt(F/N));
   model transfprop = variety;
   random intercept / subject=block;
   lsmeans variety;
   ods output lsmeans=LSM;
run;


data BackTrans;
   set LSM;
   mu=(sin(estimate))**2;
   se_mu=2*(cos(estimate))*stderr;
run;


proc print data=BackTrans;
run;
```

FIG. 5-13. GLIMMIX output containing the test for the fixed effect and parameter estimates for the fit to the binomial using the arcsine-square root transformation in Example 5.5.

Type III Tests of Fixed Effects

Effect	Num DF	Den DF	F Value	Pr > F
Variety	1	9	2.18	0.0714

Variety Least Squares Means

Variety	Estimate	Standard Error	DF	t Value	Pr > |t|
0	1.0446	0.08326	9	12.55	<0.0001
1	1.2367	0.08326	9	14.85	<0.0001

Variety Least Squares Means

Obs	Effect	Variety	Estimate	StdErr	DF	t Value	Probt	mu	se_mu
1	Variety	0	1.0446	0.08326	9	12.55	<0.0001	0.74775	0.083637
2	Variety	1	1.2367	0.08326	9	14.85	<0.0001	0.89246	0.054609

estimation and inference, but they do cloud the issue as to what, exactly, they estimate. For this reason, transformations are increasingly difficult to justify even for non-normal fixed effects models.

5.4 THREE SIMPLES EXAMPLES

The examples in this section are based on relatively simple but commonly used designs in agricultural research. The counts and proportions in the first two examples are typical of non-continuous measurements of interest to scientists. The first example considers count data from a randomized complete block design. In the second example the data are proportions from a randomized complete block design that was repeated for several years. The third example involves data on proportions from a one factor, completely randomized design with measurements taken over time. In all three examples, conditional generalized linear mixed models are used.

EXAMPLE 5.6

As part of a study of integrated strategies to manage a weed commonly known as nutsedge that is often found in bell pepper fields, Bangarwa et al. (2011, unpublished data) conducted a greenhouse experiment in which purple nutsedge tubers were sorted into four size categories based on their fresh weight. Viable tubers of each size were planted in containers that were randomly assigned to one of four tillage frequency treatments (weekly, biweekly, monthly, no tillage). Tillage was simulated by hand mixing the soil in the container with a trowel. The tubers were grown for 12 weeks under the assigned tillage treatment. The objective was to determine if repeated tillage could be used as a management strategy.

The design was a randomized complete block (RCB) with four blocks based on the location within the greenhouse and a 4×4 factorial treatment structure. Tuber size category (*weight*) and tillage frequency (*spacing*) were treated as fixed effects and blocks were a random effect. The tillage frequency levels were given as weeks between tillage with the no tillage treatment assigned a value of 12 weeks (the length of the experiment). The total number of new tubers produced from the original parent tuber in a container over the 12 week period was analyzed. The new tuber count was assumed to follow a Poisson distribution (Section 2.3).

For the Poisson distribution, the link function is the natural logarithm. Hence, the conditional mean of the new tuber count Y on the link scale is given by

$$\eta_{ijk} = g(E[Y_{ijk} \mid B_k]) = \log(E[Y_{ijk} \mid B_k]) = \beta_0 + B_k + S_i + W_j + SW_{ij} + BSW_{ijk} \text{ for } i, j, k = 1, 2, 3, 4$$

where β_0 is the overall mean, S_i is the ith tillage frequency effect, W_j is the jth parent tuber weight category effect, SW_{ij} is the tillage frequency $\times$ weight category interaction effect, B_k is the kth block effect, and BSW_{ijk} is the block $\times$ tillage frequency $\times$ weight category interaction effect.

The model can be written in matrix form as

$$\eta = \log(E[Y \mid u]) = X\beta + Zu$$

where X is the 64×25 fixed effects design matrix and Z is the 64×68 random effects design matrix. The vector β contains the 25 fixed effects parameters (overall mean + 4 tuber size effects + 4 tillage effects + 16 tillage $\times$ weight category interaction effects), and the vector u contains the 68 random effects (4 block effects + 64 block $\times$ tillage $\times$ weight category interaction effects).

As before, we assume that the distribution of the random effects is multivariate normal; i.e., $u \sim MVN(0, G)$. The covariance matrix G is a diagonal matrix with the block and block $\times$ tillage $\times$ weight category interaction variances on the diagonal. The conditional covariance matrix of Y given u (the R matrix) does not contain any additional parameters since the mean and variance of a Poisson distribution are equal.

The GLIMMIX statements used to fit the model are shown in Fig. 5–14. The *method* option on the PROC GLIMMIX statement indicates that the maximum likelihood estimates were obtained using the Laplace method. The *model* statement options specify the Poisson distribution and the natural logarithm link function. The *covtest* statement provides a test of the hypothesis that the block variance is zero. See Example 4.5 for additional details on testing covariance parameters.

Figure 5–15 contains the basic model and fitting information. Since the R matrix does not contain any additional parameters, it is not listed in *Dimensions*

FIG. 5-14. GLIMMIX statements to fit the Poisson model for Example 5.6.

```
proc glimmix data=tuber method=laplace plots=studentpanel;
   class block weight spacing;
   model total = spacing weight spacing*weight / dist=poisson link=log;
   random intercept spacing*weight / subject=block;
   covtest 'blocks' 0 . ;
   covtest 'block x sp x wt' . 0 ;
   lsmeans spacing*weight / ilink plot=meanplot(sliceby=weight join);
   lsmeans spacing*weight / ilink plot=meanplot(sliceby=weight join ilink);
   output out=new pred(ilink)=predi stderr(ilink)=sepredi pred=pred stderr=sepred
      resid=resid student=student;
run;
```

FIG. 5-15. GLIMMIX output containing the basic model and fitting information for Example 5.6.

Model Information

Data Set	WORK.TUBER
Response Variable	Total
Response Distribution	Poisson
Link Function	Log
Variance Function	Default
Variance Matrix Blocked By	Block
Estimation Technique	Maximum Likelihood
Likelihood Approximation	Laplace
Degrees of Freedom Method	Containment

Class Level Information

Class	Levels	Values
block	4	1 2 3 4
weight	4	0.25–0.50 0.50–0.75 0.75–1.00 <0.25
spacing	4	1 2 4 12

Number of Observations Read	64
Number of Observations Used	64

Dimensions

G-side Cov. Parameters	2
Columns in X	25
Columns in Z per Subject	17
Subjects (Blocks in V)	4
Max Obs per Subject	16

FIG. 5-16. GLIMMIX output containing the estimated covariance parameters and the tests of the hypothesis that the variance is zero for Example 5.6.

Covariance Parameter Estimates

Cov Parm	Subject	Estimate	Standard Error
Intercept	block	0.000405	0.002375
weight*spacing	block	4.87E-19	0.02997

Tests of Covariance Parameters Based on the Likelihood

Label	DF	−2 Log Like	ChiSq	Pr > ChiSq	Note
Blocks	1	340.11	0.11	0.7349	—
block x sp x wt	1	339.99	.	1.0000	MI

MI: P-value based on a mixture of chi-squares.
—: Standard test with unadjusted p-values.

FIG. 5-17. GLIMMIX output containing the conditional distribution fit statistics for Example 5.6.

Fit Statistics for Conditional Distribution

-2 log L(total | r. effects)	338.26
Pearson Chi-Square	46.24
Pearson Chi-Square / DF	0.72

section of the output nor is there an estimated residual variance in the *Covariance Parameter Estimates* section.

The estimated variance components are given in Fig. 5–16. The variance of the block × tillage × weight category interaction random effect is essentially zero (4.87 × 10^{-19}). Based on the results of the chi-square tests, neither variance is significantly different from zero. For greenhouse experiments where the blocks represent locations within the greenhouse, negligible block variability may not be unreasonable.

The Pearson chi-square/df provides a goodness of fit statistic to assess the assumed mean–variance relationship of the Poisson. Because the mean and variance of the Poisson are equal, the scale parameter ϕ is known to be one. If the Poisson assumption is satisfied, the Pearson chi-square/df should be close to one. Its estimated value of 0.72 in the *Fit Statistics* section (Fig. 5–17) does not indicate strong evidence of departure from the Poisson requirement.

Based on the fixed effects tests in Fig. 5–18, the tillage frequency × weight category interaction is significant. The least squares means are also listed in Fig. 5–18. The means on the link scale (natural logarithm) are listed in the *Estimate* column, followed by their estimated standard errors on the link scale in the *Standard Error* column. The *ilink* option on the *lsmeans* statement applies the inverse link function to

FIG. 5-18. GLIMMIX output containing the tests for the fixed effects and the interaction least squares means for Example 5.6.

Type III Tests of Fixed Effects

Effect	Num DF	Den DF	F Value	Pr > F
spacing	3	45	670.52	<0.0001
weight	3	45	7.57	0.0003
weight*spacing	9	45	14.06	<0.0001

weight*spacing Least Squares Means

Initial tuber fresh weight	Tillage spacing (wk)	Estimate	Standard Error	DF	t Value	Pr > |t|	Mean	Standard Error Mean
0.25-0.50	1	1.5040	0.2375	45	6.33	<0.0001	4.4997	1.0686
0.25-0.50	2	1.3862	0.2516	45	5.51	<0.0001	3.9997	1.0063
0.25-0.50	4	2.8903	0.1210	45	23.89	<0.0001	17.9983	2.1779
0.25-0.50	12	4.7916	0.04840	45	99.01	<0.0001	120.49	5.8313
0.50-0.75	1	1.5040	0.2383	45	6.31	<0.0001	4.4997	1.0723
0.50-0.75	2	1.5040	0.2371	45	6.34	<0.0001	4.4997	1.0670
0.50-0.75	4	3.4339	0.09244	45	37.15	<0.0001	30.9969	2.8654
0.50-0.75	12	4.7998	0.05111	45	93.91	<0.0001	121.49	6.2095
0.75-1.00	1	1.6581	0.2194	45	7.56	<0.0001	5.2495	1.1517
0.75-1.00	2	1.5040	0.2373	45	6.34	<0.0001	4.4996	1.0679
0.75-1.00	4	4.0900	0.06716	45	60.90	<0.0001	59.7425	4.0124
0.75-1.00	12	4.8121	0.05791	45	83.10	<0.0001	122.99	7.1221
<0.25	1	1.3862	0.2541	45	5.46	<0.0001	3.9998	1.0163
<0.25	2	1.3217	0.2674	45	4.94	<0.0001	3.7498	1.0028
<0.25	4	2.2772	0.1725	45	13.20	<0.0001	9.7492	1.6820
<0.25	12	4.7770	0.05791	45	82.49	<0.0001	118.74	6.8762

produce estimates on the data scale. These estimated counts are listed in the column labeled *Mean*. The delta method was used to obtain the approximate estimated standard errors on the count scale shown in the rightmost column of the table.

The least squares means on the log scale are graphed in Fig. 5–19 and their inverse linked values on the data scale are shown in Fig. 5–20. Note that the vertical axis label for the estimates on the data scale is denoted by *"Inverse linked..."*. Both graphs are somewhat misleading because GLIMMIX considers the tillage frequencies as labels and not as numerical values and places them at equally spaced intervals on the horizontal axis. Despite this ambiguity, it is clear from both graphs that there is an increasing trend as a function of tillage frequency that is not the same across all weight categories. The trends can be analyzed further using either of two approaches. The tillage frequencies could be treated as numerical and analyzed as a regression problem using analysis of covariance with

FIG. 5-19. GLIMMIX output displaying a graph of the least squares means for the tillage frequency × weight category interaction on the link scale (natural logarithm) in Example 5.6.

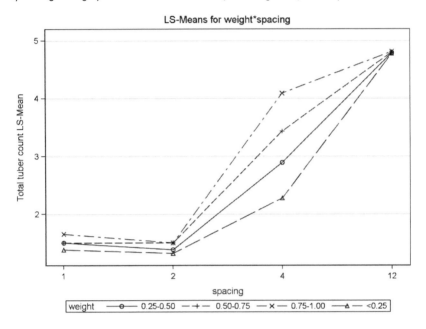

FIG. 5-20. GLIMMIX output displaying a graph of the least squares means for the tillage frequency × weight category interaction on the data scale (total tuber count) in Example 5.6.

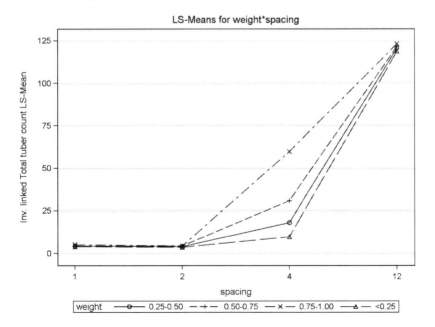

frequency as the covariate. The alternative would be to construct orthogonal polynomial trend contrasts. The choice of analysis would depend in part on whether or not actual regression equations would be required to meet the objectives of the experiment or if indications of the general shape of the trends would suffice. Since the ideas behind the regression–analysis of covariance approach were illustrated in Example 4.6, the trend contrast approach will be taken in this example.

The coefficients listed in commonly available tables of orthogonal polynomials are not applicable in this example because the tables are constructed for equally spaced treatments. However, PROC IML can be used to obtain the required coefficients for the above tillage frequency spacings. The IML statements are shown in Fig. 5–21. The *levels* statement contains the numerical values of the tillage spacings and the *maxdegree* statement calculates the maximum degree of the polynomial. The *orpol* function in the *contrast* statement creates the actual coefficients displayed in Fig. 5–22.

FIG. 5-21. IML statements to obtain the coefficients for the orthogonal polynomial trend contrasts for Example 5.6.

```
proc iml;
    reset print nolog name;
    Levels = { 1  2  4  12 };
    Weights = { 1  1  1  1 };
    MaxDegree = ncol(levels) - 1;
    mattrib Contrasts rowname = {'Constant' 'Linear' 'Quadratic' 'Cubic'};
    contrasts = t(orpol(Levels, MaxDegree, Weights));
run;
```

FIG. 5-22. IML output containing the orthogonal polynomial trend contrast coefficients for Example 5.6.

Levels (numeric) 1 row 4 cols

1	2	4	12

Weights (numeric) 1 row 4 cols

1	1	1	1

MaxDegree (numeric) 1 row 1 col

3

Contrasts (numeric) 4 rows 4 cols

Constant	0.5	0.5	0.5	0.5
Linear	−0.433736	−0.318073	−0.086747	0.8385567
Quadratic	0.5688432	0.0091749	−0.793628	0.2156099
Cubic	−0.48815	0.8054472	−0.335603	0.0183056

The *contrast* statements are shown in Fig. 5–23. For the interaction contrasts, the first set of coefficients compares the first and second weight categories, the second set compares the second and third weight categories, and the last set compares the third and fourth weight categories. Based on the results in Fig. 5–24, the trends in the least squares means on the log scale depend on the weight category and are more complex than a simple quadratic model since the cubic main effect trend contrast is significant.

The panel of conditional studentized residual plots is shown in Fig. 5–25. The 46th observation in the data set (block = 2, weight category < 0.25, tillage frequency = monthly, tuber count = 5) has a conditional studentized residual of −3.48. Its effect on the results could be explored by removing it from the data file and refitting the model. In SAS (SAS Institute, Cary, NC), not removing the observation from the data file but setting the response (tuber count) to missing will give a predicted value from that observation while not using it in the model fitting process. Comparing the predicted values with and without the observation in the analysis can sometimes offer additional insight about the data to the researcher. ■

FIG. 5–23. GLIMMIX statements to test the orthogonal polynomial main effect and interaction trend contrasts for Example 5.6.

```
proc glimmix data=tuber method=laplace plots=studentpanel;
   class block weight spacing;
   model total = spacing weight spacing*weight / dist=poisson link=log;
   random intercept spacing*weight / subject=block;

   contrast 'Linear spacing trend'    spacing  -0.4337360 -0.3180730 -0.086747 0.8385567;
   contrast 'Quadratic spacing trend'  spacing  0.5688432 0.0091749 -0.793628 0.2156099;
   contrast 'Cubic spacing trend'      spacing  -0.4881500 0.8054472 -0.335603 0.0183056;

   contrast 'Linear spacing*weight'  spacing*weight -0.4337360 -0.3180730 -0.086747
      0.8385567 0.4337360 0.3180730 0.086747 -0.8385567 0 0 0 0 0 0 0 0,
      spacing*weight 0 0 0 0 -0.4337360 -0.3180730 -0.086747 0.8385567 0.4337360
      0.3180730 0.086747 -0.8385567 0 0 0 0,
      spacing*weight 0 0 0 0 0 0 0 0 -0.4337360 -0.3180730 -0.086747 0.8385567 0.4337360
      0.3180730 0.086747 -0.8385567;
   contrast 'Quadratic spacing*weight'  spacing*weight 0.5688432 0.0091749 -0.793628
      0.2156099 -0.5688432 -0.0091749 0.793628 -0.2156099 0 0 0 0 0 0 0 0,
      spacing*weight 0 0 0 0 0.5688432 0.0091749 -0.793628 0.2156099 -0.5688432
      -0.0091749 0.793628 -0.2156099 0 0 0 0,
      spacing*weight 0 0 0 0 0 0 0 0 0.5688432 0.0091749 -0.793628 0.2156099 -0.5688432
      -0.0091749 0.793628 -0.2156099;
   contrast 'Cubic spacing*weight'  spacing*weight -0.4881500 0.8054472 -0.335603 0.0183056
      0.4881500 -0.8054472 0.335603 -0.0183056 0 0 0 0 0 0 0 0,
      spacing*weight 0 0 0 0 -0.4881500 0.8054472 -0.335603 0.0183056 0.4881500
      -0.8054472 0.335603 -0.0183056 0 0 0 0,
      spacing*weight 0 0 0 0 0 0 0 0 -0.4881500 0.8054472 -0.335603 0.0183056 0.4881500
      -0.8054472 0.335603 -0.0183056;

run;
```

FIG. 5-24. GLIMMIX output containing the tests for the orthogonal polynomial trend contrasts for Example 5.6.

Contrasts				
Label	Num DF	Den DF	F Value	Pr > F
Linear spacing trend	1	45	1518.64	<0.0001
Quadratic spacing trend	1	45	57.59	<0.0001
Cubic spacing trend	1	45	23.71	<0.0001
Linear spacing*weight	3	45	0.98	0.4107
Quadratic spacing*weight	3	45	10.92	<0.0001
Cubic spacing*weight	3	45	1.11	0.3563

FIG. 5-25. GLIMMIX output containing graphs of the conditional studentized residuals for Example 5.6.

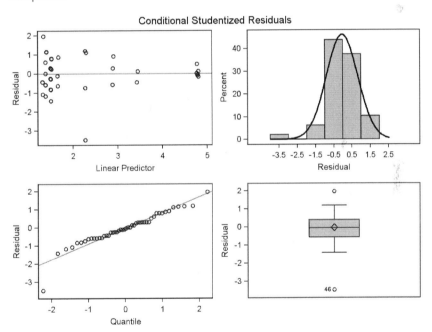

EXAMPLE 5.7

Mummy berry is an important disease of blueberry that is caused by a fungus. The initial stage of the disease manifests itself as blighted shoots on the plant. The blighted tissue produces conidia that are dispersed by bees to the flowers during pollination, eventually resulting in infected fruit. The infected fruit fall to the ground and serve as the overwintering stage for the next year's infection.

From a long-term study reported by Ehlenfeldt et al. (2010), six blueberry cultivars from a larger set that were tested yearly, with the exception of 2004, from 1995 through 2007 will be analyzed in this example. Two of the cultivars were not tested in 2006, and one was not tested in 2007. Tests were conducted outside in a cold frame using individually potted plants. Blighted plants from susceptible cultivars were included as an inoculum source. Berries were picked before reaching maturity and individually examined for infection. Since flowers are pollinated individually, the number of infected berries was assumed to have a binomial distribution (Section 2.3). The incidence of infection was expressed as a proportion of the total fruit on the plant. The primary objectives were to estimate the year to year variation and to determine if cultivars responded similarly to the random year effect.

The design for each year was a randomized complete block with five to eight blocks, depending on the year. The experimental unit was a single potted plant. Cultivar (*cult*) was assumed to be a fixed effect. Preliminary plots of the data versus time indicated yearly variation but no obvious time trend. Hence, year (*year*) and the year × cultivar interaction will be treated as random effects. Blocks (*block*) within years are also random.

For the binomial distribution the default link function is the logit, i.e., $\log[\pi/(1 - \pi)]$, where π is the probability of an infected berry. The conditional mean of the proportion of infected berries on a plant, P, on the link scale is given by

$$\eta_{ijk} = g(E[P_{ijk} \mid B_{k(i)}, S_i, SC_{ij}] = \text{logit}(P_{ijk} \mid B_{k(i)}, S_i, SC_{ij}) = \beta_0 + B_{k(i)} + S_i + C_j + SC_{ij'}$$

for $i = 1, ..., 12; j = 1, ..., 6; k = 1, ..., K_i$

where $K_i = 5$ or 8 and β_0 is the overall mean, S_i is the effect of the ith year (season), C_j is the effect of the jth cultivar, SC_{ij} is the effect of the year × cultivar interaction, and $B_{k(i)}$ is the effect of the kth block within the ith year.

The matrix form of the model can be written as

$$\eta = \text{logit}(E[P \mid u]) = X\beta + Zu$$

where X is the 365 × 7 fixed effects design matrix and Z is the 365 × 15 random effects design matrix. The vector β contains the 7 fixed effects parameters (overall mean + 6 cultivar effects), and the vector u contains the 15 random effects (1 year effect + 8 blocks within year effects + 6 yr × cultivar interaction effects). The G matrix contains the year, block within year, and year × cultivar interaction variances. As in Example 5.6, the R matrix does not contain any additional parameters.

The GLIMMIX statements used to fit the model are shown in Fig. 5–26. As in Example 5.1, the sample proportion was expressed on the *model* statement as a ratio of the variables representing the number of infected berries (*infect*) and the total number of berries on the plant (*total*). The binomial distribution and logit link are also specified on the *model* statement.

Figure 5–27 contains the basic model and fitting information. The output indicates that data were obtained from 365 plants over the course of the experiment. There were a total of 9458 infected berries out of 42,058 berries on the plants. Each block in the covariance matrix of the estimated proportions, V, corresponds to one

FIG. 5-26. GLIMMIX statements to fit the initial binomial logit model for Example 5.7.

```
proc glimmix data=berry method=laplace plots=(studentpanel);
  class year block cult;
  model infect/total = cult / dist=bin link=logit;
  random intercept block / subject=year G;
  random intercept / subject=year*cult G;
  covtest 'Year'            0 . . ;
  covtest 'Block within year' . 0 . ;
  covtest 'Year*cultivar'    . . 0 ;
  lsmeans cult / ilink cl adjust=tukey lines plot=(diffplot(noabs center));
  output out=new pred=predlogit pred(ilink)=predprop stderr=selogit stderr(ilink)=seprop
    resid=residlogit resid(ilink)=residprop student=student;
run;
```

FIG. 5-27. GLIMMIX output containing the basic model and fitting information for Example 5.7.

Model Information

Data Set	WORK.BERRY
Response Variable (Events)	infect
Response Variable (Trials)	total
Response Distribution	Binomial
Link Function	Logit
Variance Function	Default
Variance Matrix Blocked By	year
Estimation Technique	Maximum Likelihood
Likelihood Approximation	Laplace
Degrees of Freedom Method	Containment

Class Level Information

Class	Levels	Values
year	12	1995 1996 1997 1998 1999 2000 2001 2002 2003 2005 2006 2007
block	8	1 2 3 4 5 6 7 8
cult	6	ATL BJA BRA NOS RAN SIE

Number of Observations Read	365
Number of Observations Used	365
Number of Events	9458
Number of Trials	42058

Dimensions

G-side Cov. Parameters	3
Columns in X	7
Columns in Z per Subject	15
Subjects (Blocks in V)	12
Max Obs per Subject	48

of the 12 years. Each of these blocks consists of two sub-blocks, one generated by each of the *random* statements. The G option on the *random* statements instructs GLIMMIX to print the blocks for the first year.

The estimated variance components are shown in Fig. 5–28 along with the tests for each variance component equal to zero from the corresponding *covtest* statement. All three variances are highly significant ($p < 0.0001$). The significant year × cultivar interaction variance demonstrates that the cultivars do not respond to the random year effect in the same way. Approximately 60% of the total variance on the logit scale $[1.1213/(1.1213 + 0.1912 + 0.5412)]$ can be attributed to the year effect and only 10% to the blocks within year effect. Since blocking was used to ensure that the cultivars were spread somewhat uniformly within the confines of the cold frame, it is not unreasonable that its variance was small relative to the year variance. Approximately 30% of the variance can be attributed to cultivars responding inconsistently across years.

The result of the fixed effects test in Fig. 5–29 indicates that the probability of infection among cultivars differs on the logit scale. The least squares means and their pairwise differences based on the Tukey–Kramer procedure are shown in Fig. 5–30. The estimated means on the logit scale are listed in the *Estimate* column. The

FIG. 5–28. GLIMMIX output containing the estimated covariance parameters and the tests of the hypothesis that the variance is zero for Example 5.7.

Covariance Parameter Estimates

Cov Parm	Subject	Estimate	Standard Error
Intercept	Year	1.1213	0.5127
block	Year	0.1912	0.04315
Intercept	year*cult	0.5412	0.1149

Tests of Covariance Parameters Based on the Likelihood

Label	DF	−2 Log Like	ChiSq	Pr > ChiSq	Note
Year	1	3194.83	41.04	<0.0001	MI
Block within year	1	3535.12	381.33	<0.0001	MI
Year*cultivar	1	4223.68	1069.90	<0.0001	MI

MI: P-value based on a mixture of chi-squares.

FIG. 5–29. GLIMMIX output containing the results of the fixed effects test for Example 5.7.

Type III Tests of Fixed Effects

Effect	Num DF	Den DF	F Value	Pr > F
cult	5	52	47.21	<0.0001

FIG. 5–30. GLIMMIX output containing the least squares means for the cultivar effect and the pairwise differences at $\alpha = 0.05$ for Example 5.7.

cult Least Squares Means

Blueberry cultivar	Estimate	Standard Error	DF	t Value	Pr > |t|	Lower	Upper	Mean	Standard Error Mean	Lower Mean	Upper Mean
ATL	0.2469	0.3874	52	0.64	0.5267	-0.5305	1.0243	0.5614	0.09539	0.3704	0.7358
BJA	-2.9662	0.3908	52	-7.59	<0.0001	-3.7505	-2.1819	0.04898	0.01821	0.02297	0.1014
BRA	-0.9777	0.3786	52	-2.58	0.0127	-1.7373	-0.2180	0.2734	0.07520	0.1497	0.4457
NOS	-4.8486	0.4363	52	-11.11	<0.0001	-5.7240	-3.9731	0.007779	0.003367	0.003256	0.01847
RAN	-1.6677	0.3801	52	-4.39	<0.0001	-2.4304	-0.9050	0.1587	0.05076	0.08088	0.2880
SIE	-0.5664	0.3785	52	-1.50	0.1406	-1.3260	0.1932	0.3621	0.08743	0.2098	0.5481

Differences of cult Least Squares Means
Adjustment for Multiple Comparisons: Tukey-Kramer

Blueberry cultivar	Blueberry cultivar	Estimate	Standard Error	DF	t Value	Pr > |t|	Adj P	Lower	Upper	Adj Lower	Adj Upper
ATL	BJA	3.2131	0.3317	52	9.69	<0.0001	<0.0001	2.5475	3.8786	2.2318	4.1944
ATL	BRA	1.2246	0.3167	52	3.87	0.0003	0.0040	0.5891	1.8601	0.2876	2.1615
ATL	NOS	5.0955	0.3853	52	13.22	<0.0001	<0.0001	4.3222	5.8687	3.9554	6.2355
ATL	RAN	1.9146	0.3188	52	6.01	<0.0001	<0.0001	1.2749	2.5543	0.9714	2.8577
ATL	SIE	0.8133	0.3170	52	2.57	0.0132	0.1243	0.1773	1.4493	-0.1244	1.7511
BJA	BRA	-1.9885	0.3210	52	-6.19	<0.0001	<0.0001	-2.6326	-1.3444	-2.9382	-1.0388
BJA	NOS	1.8824	0.3839	52	4.90	<0.0001	0.0001	1.1121	2.6527	0.7467	3.0181
BJA	RAN	-1.2985	0.3227	52	-4.02	0.0002	0.0024	-1.9461	-0.6509	-2.2533	-0.3437
BJA	SIE	-2.3998	0.3209	52	-7.48	<0.0001	<0.0001	-3.0437	-1.7558	-3.3493	-1.4502
BRA	NOS	3.8709	0.3748	52	10.33	<0.0001	<0.0001	3.1187	4.6230	2.7619	4.9799
BRA	RAN	0.6900	0.3077	52	2.24	0.0292	0.2367	0.07250	1.3075	-0.2204	1.6004
BRA	SIE	-0.4113	0.3058	52	-1.34	0.1845	0.7587	-1.0248	0.2023	-1.3159	0.4934
NOS	RAN	-3.1809	0.3757	52	-8.47	<0.0001	<0.0001	-3.9348	-2.4270	-4.2924	-2.0694
NOS	SIE	-4.2821	0.3747	52	-11.43	<0.0001	<0.0001	-5.0341	-3.5302	-5.3908	-3.1735
RAN	SIE	-1.1013	0.3077	52	-3.58	0.0008	0.0094	-1.7187	-0.4838	-2.0116	-0.1909

t-test for the null hypothesis that the mean on the logit scale is zero is equivalent to testing the hypothesis that the probability of an infected berry is 0.50. The back-transformed estimated probabilities of infection are given in the *Mean* column along with their estimated standard errors calculated by the delta method (Section 3.2). The 95% confidence intervals for the estimated probabilities were obtained by back-transforming the confidence interval endpoints for the estimated logit. The Tukey–Kramer *p*-values for the pairwise comparisons on the logit scale are given in the *Adj P* column of the table of differences.

The *lines* option on the *lsmeans* statement produced the table of estimated means on the logit scale shown in Fig. 5–31. This option only applies to main effects of fixed factors. If the variances of the means are not equal as in the present case, the *lines* option may detect fewer significant differences than the t-tests for the individual comparisons in Fig. 5–30. If this occurs, the differences in the results will be noted on the output. In this example, the sets of differences are the same in both figures.

FIG. 5-31. GLIMMIX output containing the table of least squares means created by the *lines* option on the *lsmeans* statement for Example 5.7.

**Tukey-Kramer Grouping
for cult Least Squares Means
(Alpha=0.05)**
LS-means with the same letter are
not significantly different.

Blueberry cultivar	Estimate		
ATL	0.2469		A
			A
SIE	-0.5664	B	A
		B	
BRA	-0.9777	B	C
			C
RAN	-1.6677		C
BJA	-2.9662		D
NOS	-4.8486		E

FIG. 5-32. GLIMMIX output containing the conditional distribution fit statistics for Example 5.7.

Fit Statistics for Conditional Distribution	
-2 log L(infect | r. effects)	2669.41
Pearson Chi-Square	1399.05
Pearson Chi-Square / DF	3.83

The Pearson chi-square/df estimate of 3.83 in Fig. 5–32 indicates that there may be some over-dispersion in the data. Over-dispersion would imply more variability in the data than would be expected under the binomial model. The studentized residual versus the linear predictor graph in Fig. 5–33 provides additional evidence of possible over-dispersion. There are 28 studentized residuals, approximately 7.7% of the observations, with absolute values greater than 4. Since the berries grow in clusters on the plant, over-dispersion may be indicative of berries within clusters not being independent. Over-dispersion will be discussed in more detail in Sections 5.5 through 5.7. ■

EXAMPLE 5.8

Urea loses nitrogen through volatilization once it is applied to the soil. There are a number of commercial products that claim to reduce nitrogen loss. Data for

FIG. 5-33. GLIMMIX output containing graphs of the conditional studentized residuals in Example 5.7.

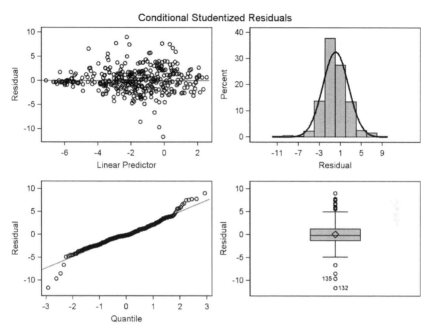

this example (used with permission) are from a series of laboratory experiments conducted by T.L. Roberts to compare the effectiveness of such products. This particular experiment included three such products. The products were assigned completely at random to samples from the same soil, three soil samples per product. The cumulative percentage of applied nitrogen lost through volatilization in the form of ammonia was measured on each sample at 3, 7, 11, and 15 days after application of the urea to the soil.

The response variable is a cumulative proportion that varies between 0 and 1. In this sense, it is similar to the probability of a success in a binomial distribution. However, the volatilization rate is clearly not a binomial response; that is, it is not the number of successes out of a finite number of trials. Instead, the volatilization rate is a continuous variable. A beta distributed random variable is continuous, varies between 0 and 1, and has a probability function that is flexible for assuming different shapes (Section 2.3). We will assume that the cumulative proportion has a beta distribution. The canonical link function for the beta distribution is the logit, i.e., $\log[\mu/(1 - \mu)]$, where μ is the mean of the beta distribution.

One approach to the analysis of these data is to treat the experiment as a one factor, completely randomized design with time as a repeated measure. The conditional mean of the cumulative proportion of lost nitrogen, Y, would be expressed as

$$\eta_{ijk} = \log\left(\frac{\mu_{ijk}}{1-\mu_{ijk}}\right) = \beta_0 + P_i + T_j + PT_{ij} + R_{k(i)}$$

where μ_{ijk} = E$[Y_{ijk} \mid R_{k(i)}]$ is the conditional mean of Y, β_0 is the overall mean, P_i is the ith product effect, T_j is the jth time effect, PT$_{ij}$ is the product × time interaction effect, and $R_{k(i)}$ is the kth replication effect for the ith product. This model treats time as a categorical variable.

Alternatively, it may be more informative to model the time and product × time effects using polynomial regression that treats time as a continuous variable. The regression form of the linear predictor would be given by

$$\eta_{ijk} = \log\left(\frac{\mu_{ijk}}{1-\mu_{ijk}}\right) = \beta_0 + P_i + \beta_{1i}D_j + \beta_{2i}D_j^2 + \beta_{3i}D_j^3 + R_{k(i)}$$

where β_{1i}, β_{2i}, and β_{3i} are the linear, quadratic, and cubic regression coefficients, respectively, for the ith product, and D_j is the number of days after application at the jth sampling time. The constant terms $\beta_0 + P_i$ could also be expressed as β_{0i} or alternatively, each of the regression coefficients β_{1i}, β_{2i}, and β_{3i} could be expressed as the sum of an overall constant and a product effect. The cubic and quadratic terms may be removed from the model if justified statistically as the analysis proceeds.

The data were analyzed initially as a repeated measures design using the methods discussed in Section 4.5 and which will be illustrated again in Section 5.8. The details of that analysis will not be shown here. Suffice it to say that plausible covariance models showed no evidence of serial correlation. Unless the analysis shows otherwise, repeated observations on an experimental unit should be assumed to be correlated with the correlation decreasing as the observations become farther apart in time. In this example, the apparent lack of serial correlation may be due to the length of the time (4 days) between observations on a sample.

As indicated above, the data could also be analyzed using a regression model approach with time treated as a quantitative variable. This approach will be taken for the remainder of this example. A plot of the means over the replications can be used to obtain an indication of the shape of the regression functions and, hence, what might be expected as a reasonable model. These plots can be obtained easily by treating time as a qualitative variable, fitting the repeated measures model, and plotting the least squares means. The GLIMMIX statements to produce these plots are given in Fig. 5–34, and the graphs are shown in Fig. 5–35 (link scale) and 5–36 (data scale).

Since the analysis is performed on the link scale, Fig. 5–35 provides more realistic guidance for proceeding with the analysis. In both figures, changes over time appear to be curvilinear. The response profiles for products 1 and 2 may be quadratic, but product 3 may require a cubic model. Differences between the products are noticeable, especially between product 1 and the other two products, which appear to be similar. Visually there appears to be an interaction between the products and time; that is, the rate of change over time differs by product. Note

FIG. 5-34. GLIMMIX statements to fit the repeated measures with independent errors model to determine the form of the regression function in Example 5.8.

```
proc glimmix data=ammonia method=laplace plots=studentpanel;
    class product rep time;
    model cumprop = product | time / dist=beta link=logit;
    random intercept / subject=rep(product);
    lsmeans product*day / plot=meanplot(sliceby=product join);
    lsmeans product*day / plot=meanplot(sliceby=product join ilink);
run;
```

FIG. 5-35. GLIMMIX output containing the product × time interaction means on the logit scale in Example 5.8.

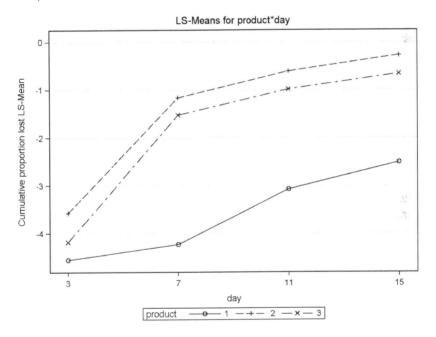

that interaction can depend on the scale. On the data scale (Fig. 5–36), the rate of change for product 1 is small and different from products 2 and 3, whereas on the link scale a large rate of change for product 1 occurs between Days 7 and 11 while the rates for products 2 and 3 behave similarly on the link scale.

Figure 5–37 shows the GLIMMIX statements to fit the cubic polynomial model with coefficients depending on the products. The time effect is denoted by *day* rather than by *time* as in Fig. 5–34 to reinforce its role as a quantitative and not a qualitative predictor. Not including *day* in the *class* statement informs GLIMMIX that it is quantitative. The vertical bar notation in the *model* statement indicates that all possible products of the listed terms should be included in the fitted model.

FIG. 5-36. GLIMMIX output containing the product × time interaction means on the cumulative proportion (data) scale in Example 5.8.

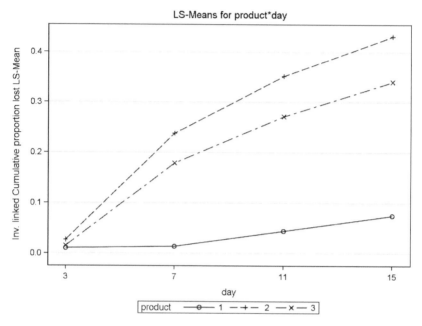

FIG. 5-37. GLIMMIX statements to fit the cubic polynomial regression model with coefficients depending on product for Example 5.8.

```
proc glimmix data=ammonia2  method=laplace  plots=studentpanel;
  class product rep;
  model cumprop = product | day | day | day / dist=beta link=logit;
  random intercept / subject=rep(product);
  output out=new1  pred=pred stderr=sepred resid=resid student=student;
run;
```

This short hand notation generates the various powers of *day* and their interactions with *product*.

The basic model and fitting information as well as portions of the *Iteration History* section of the output are shown in Fig. 5–38. The 16 columns in the fixed effects design matrix **X** correspond to an overall average coefficient and three product effects that make up the intercepts, linear, quadratic, and cubic regression coefficients. The covariance parameter estimates are for the variance among replications and the scale parameter ϕ of the beta distribution, respectively. The nine subjects are the three replications for each of the products.

The covariance parameter estimates are shown in Fig. 5–39. The results in this figure and the SAS program log contain features that should be considered

FIG. 5-38. GLIMMIX output containing basic model and fitting information for the cubic polynomial regression model with coefficients depending on product in Example 5.8.

Model Information

Data Set	WORK.AMMONIA2
Response Variable	cumprop
Response Distribution	Beta
Link Function	Logit
Variance Function	Default
Variance Matrix Blocked By	rep(product)
Estimation Technique	Maximum Likelihood
Likelihood Approximation	Laplace
Degrees of Freedom Method	Containment

Class Level Information

Class	Levels	Values
product	3	1 2 3
rep	3	1 2 3

Number of Observations Read	36
Number of Observations Used	36

Dimensions

G-side Cov. Parameters	1
R-side Cov. Parameters	1
Columns in X	16
Columns in Z per Subject	1
Subjects (Blocks in V)	9
Max Obs per Subject	4

Iteration History

Iteration	Restarts	Evaluations	Objective Function	Change	Max Gradient
0	0	4	−226.6691193	.	5208.38
1	0	9	−226.6705944	0.00147516	605.5484
2	0	3	−226.6710157	0.00042123	234.5629
3	0	3	−226.6710501	0.00003439	272.4434
. . . .					
18	0	2	−227.1593416	0.00056654	62.71895
19	0	3	−227.1594728	0.00013124	6.041157
20	0	6	−227.1595985	0.00012566	562.6571

red flags before proceeding further with interpretation of the estimates and test statistics. Note that the estimated standard error for the scale parameter is blank; i.e., a dot (.) appears. The SAS log contains a warning that "at least one element of gradient is greater than 10^{-3}." This can also be seen in the last line of the *Max Gradient* column of the *Iteration History*. A maximum gradient that is not close to zero is an indication that the optimization procedure has not found a maximum

FIG. 5-39. GLIMMIX output containing the covariance parameter and scale parameter estimates for the cubic polynomial regression model with coefficients depending on product in Example 5.8.

Covariance Parameter Estimates

Cov Parm	Subject	Estimate	Standard Error
Intercept	rep(product)	0.005488	0.01040
Scale		531.70	.

FIG. 5-40. GLIMMIX output containing the covariance parameter and scale parameter estimates for the ANOVA model in Fig. 5-34 including the *nloptions* statement for Example 5.8.

Covariance Parameter Estimates

Cov Parm	Subject	Estimate	Standard Error
product	rep	0.006884	.
Scale		1010.58	268.90

of the likelihood. A standard tactic to resolve these issues is to use an alternative optimization procedure that is better suited to certain generalized linear mixed models—this particular example being one such model. Two standard alternative algorithms available in SAS can be invoked using an *nloptions* statement. These procedures can be applied by adding the statement

nloptions tech = nrridg;

or

nloptions tech = newrap;

immediately after the *random* statement in Fig. 5-37.

Unfortunately in this case, both attempts result in a failure of the procedure to converge. This sometimes happens as an artifact of fitting the regression model. When this happens, treating time as a qualitative factor and fitting the ANOVA model often helps. This is accomplished by adding either of the above *nloptions* statements immediately after the *random* statement in Fig. 5-34 and adding the time variable to the *class* statement. Using either of the *nloptions* statements yields an acceptable maximum gradient (5.62×10^{-6}). However, as shown in Fig. 5-40, the standard error for the variance among replications is missing.

After exhausting reasonable efforts to find a computing option that would produce estimates of the covariance parameters and their standard errors, the next step is to assess the impact of the missing standard error using a simulation. This was done using the method described in Section 7.4. We generated 1000 simulated experiments according to the beta ANOVA model with mean, variance, and scale parameters equal to the values based on the estimates in Fig. 5–40. Using the *nrridg* option, missing standard errors occurred for all 1000 estimates of the replication variance. However, the gradient was an issue in only 23 of the 1000 simulated data sets (compared to nearly all of the experiments using the default algorithm). The estimates of all treatment combination means on the data scale were accurate (whereas they were not when using the default). The estimated proportion for each treatment combination was essentially equal to the value set in the simulation to generate the data; their standard errors tended to slightly underestimate the standard deviation of the observed sampling distribution. This might result in inflated test statistics, although compared to an analysis using normal theory linear mixed models assuming the proportions are normally distributed, the F values using the beta generalized linear mixed model were considerably lower and, hence, more conservative than the normal theory linear mixed model.

Thus, these data illustrate a case for which there is a red flag that should be checked. The appropriate way to check the red flag that cannot be removed using a different algorithm is via simulation. If the simulation indicates that the results cannot be trusted, another model must be used. If, as was the case here, the simulation indicates that, red flag notwithstanding, the results can be trusted, one can proceed with the analysis and interpretation.

As a final comment, over the decade after PROC MIXED was first released by SAS until the release of Version 9, the computational algorithms went through several refinements. Issues that appeared in the early releases were eventually resolved, so that computational issues for normal theory linear mixed models now are rare, assuming the model is not egregiously misspecified. PROC GLIMMIX was released in 2005 and was significantly refined with the release of Version 9.2 in 2008. In other words, GLIMMIX is now in the midst of the same refining process that PROC MIXED went through in the 1990s. This data set illustrates an example of what one may encounter in the current version and how to deal with it.

To proceed with the analysis of the regression model, use the variance component estimates in Fig. 5–40 and the *hold* option in the *parms* statement to prevent GLIMMIX from attempting to re-estimate the variance components. The GLIMMIX statements are shown in Fig. 5–41. The numerical values in parentheses in the *parms* statement are the estimates from Fig. 5–40 in the corresponding

FIG. 5–41. GLIMMIX statements to fit the cubic polynomial regression model with coefficients depending on product for Example 5.8 while holding the covariance parameter estimates fixed at the values in Fig. 5–40.

```
proc glimmix data=ammonia2 method=laplace plots=studentpanel;
  class product rep;
  model cumprop = product | day | day | day / dist=beta link=logit;
  random intercept / subject=rep(product);
  parms (0.006884) (1010.58) / hold=1,2;
run;
```

FIG. 5-42. GLIMMIX output containing the covariance parameter values held fixed by the *parms* statement in Fig. 5–41 for Example 5.8.

Parameter Search

CovP1	CovP2	Objective Function
0.006884	1010.58	–231.7937222

FIG. 5-43. GLIMMIX output containing the covariance parameter values held fixed and fixed effects tests for Example 5.8.

Covariance Parameter Estimates

Cov Parm	Subject	Estimate	Standard Error
Intercept	rep(product)	0.006884	.
Scale		1010.58	.

Type III Tests of Fixed Effects

Effect	Num DF	Den DF	F Value	Pr > F
product	2	6	15.80	0.0041
day	1	18	55.61	<0.0001
day*product	2	18	24.50	<0.0001
day*day	1	18	20.56	0.0003
day*day*product	2	18	19.34	<0.0001
day*day*day	1	18	11.41	0.0034
day*day*day*product	2	18	15.98	0.0001

order. The *hold* option indicates which covariance parameters should be assumed as known values and not estimated in the model fitting process.

The effect of the *parms* statement is shown in the output presented in Fig. 5–42 and is reproduced in Fig. 5–43, which contains the covariance parameter estimates and the tests for the fixed effects regression coefficients. The small *p*-values for all of the interaction terms involving product and day indicate that the linear, quadratic, and cubic regression coefficients differ by product.

The estimated regression coefficients can be obtained using the GLIMMIX statements in Fig. 5–44. Although not required, the quadratic and cubic terms have been rewritten. For example, the quadratic term could also have been expressed as *prod*day*day* on the *model* statement. Using the *noint* and *solution* options eliminates the need to write *estimate* statements to obtain each of the 12 regression coefficients. The first set of *contrast* statements compares the intercepts across products. Similar statements could have been included for the remaining coefficients. The second set of *contrast* statements compares the linear, quadratic, and cubic coefficients simultaneously across products. A nonsignificant *p*-value for these contrasts would

FIG. 5-44. GLIMMIX statements to final fit the cubic polynomial regression model with coefficients depending on product and covariance parameters held fixed for Example 5.8

```
data ammonia3;
  set ammonia2;
  day_sq=day*day;
  day_cub=day*day_sq;
run;


proc glimmix data=ammonia3 method=laplace plots=studentpanel;
  class product rep;
  model cumprop = product day*product day_sq*product day_cub*product / noint solution
    dist=beta link=logit;
  random intercept / subject=rep(product);
  parms (0.006884)(1010.58) / hold=1,2;
  contrast 'Cubic coeff product 1 vs 2' day_cub*product 1 -1  0;
  contrast 'Cubic coeff product 1 vs 3' day_cub*product 1  0 -1;
  contrast 'Cubic coeff product 2 vs 3' day_cub*product 0  1 -1;
  contrast 'All coeff, product 1 vs 2'  day*product      1 -1  0,
                                        day_sq*product   1 -1  0,
                                        day_cub*product  1 -1  0;
  contrast 'All coeff, product 1 vs 3'  day*product      1  0 -1,
                                        day_sq*product   1  0 -1,
                                        day_cub*product  1  0 -1;
  contrast 'All coeff, product 2 vs 3'  day*product      0  1 -1,
                                        day_sq*product   0  1 -1,
                                        day_cub*product  0  1 -1;
run;
```

indicate that the regression function for the two products would differ at most by their intercepts; i.e, at most by an overall level of the cumulative proportions.

The estimated coefficients are shown in Fig. 5–45 and the results of the contrast tests in Fig. 5–46. From the second set of results in Fig. 5–46, we conclude that products 2 and 3 differ at most in their intercepts, and both products' regression functions differ significantly from that of product 1. From the first set of contrasts, the intercepts for products 2 and 3 are not significantly different. Hence, products 2 and 3 do not behave differently.

The estimates in Fig. 5–45 were used to graph the regression functions on the logit (link) scale (Fig. 5–47). The associated fitted curves for the cumulative proportion of lost nitrogen as a function of time (i.e., on the data scale) were obtained by applying the inverse link to the predicted values on the logit scale. Both graphs show differences in the effectiveness over time of the products. The predicted values for the products can be compared at selected times using the techniques described in Example 4.6. ∎

FIG. 5-45. GLIMMIX output containing the fixed covariance parameter values and the regression coefficient estimates for the final fitted model for Example 5.8.

Covariance Parameter Estimates

Cov Parm	Subject	Estimate	Standard Error
Intercept	rep(product)	0.006884	.
Scale		1010.58	.

Solutions for Fixed Effects

Effect	Product	Estimate	Standard Error	DF	t Value	Pr > |t|
product	1	−3.3527	0.8044	6	−4.17	0.0059
product	2	−7.6767	0.3967	6	−19.35	<0.0001
product	3	−8.6497	0.4853	6	−17.82	<0.0001
day*product	1	−0.6596	0.3420	18	−1.93	0.0697
day*product	2	1.7722	0.1409	18	12.57	<0.0001
day*product	3	1.9614	0.1672	18	11.73	<0.0001
day_sq*product	1	0.1021	0.04034	18	2.53	0.0210
day_sq*product	2	−0.1508	0.01570	18	−9.61	<0.0001
day_sq*product	3	−0.1692	0.01818	18	−9.31	<0.0001
day_cub*product	1	−0.00362	0.001407	18	−2.57	0.0191
day_cub*product	2	0.004370	0.000541	18	8.07	<0.0001
day_cub*product	3	0.004933	0.000616	18	8.01	<0.0001

FIG. 5-46. GLIMMIX output containing the contrast tests for equal coefficients between pairs of products for the final fitted model for Example 5.8.

Contrasts

Label	Num DF	Den DF	F Value	Pr > F
Intercept product 1 vs 2	1	6	23.25	0.0029
Intercept product 1 vs 3	1	6	31.79	0.0013
Intercept product 2 vs 3	1	6	2.41	0.1716
All coeff, product 1 vs 2	3	18	23.60	<0.0001
All coeff, product 1 vs 3	3	18	24.94	<0.0001
All coeff, product 2 vs 3	3	18	0.35	0.7922

FIG. 5-47. Graphs of the fitted cubic polynomials from the final model on the logit scale (top) and the cumulative proportion scale (bottom) for Example 5.8.

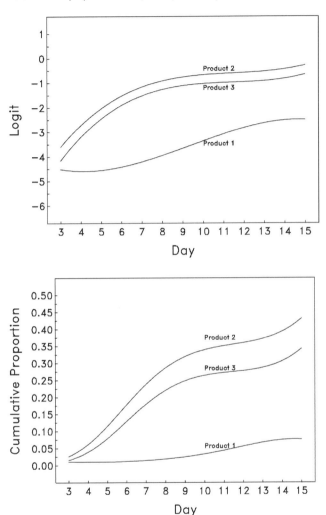

5.5 OVER-DISPERSION IN GENERALIZED LINEAR MIXED MODELS

The term over-dispersion refers to more variation displayed by data than would be expected under an assumed model. For example, if observed data are counts and are assumed to have a Poisson distribution, then, in theory, the population mean and variance should be equal. However, if the sample variance is much greater than the sample mean, then the data are said to be over-dispersed; that is, the observed variance is implausibly large for the Poisson assumption to be correct. Diagnostics such as goodness of fit statistics can be computed from the data to assess over-dispersion.

Over-dispersion is a problem that should not be ignored in an analysis. The primary, and most serious, consequence of over-dispersion is its impact on standard errors and test statistics. This was illustrated in a simple version of this problem in Example 5.1 if one failed to include the block × variety random effect. Uncorrected analysis of over-dispersed data results in underestimated standard errors, leading to confidence intervals with less than nominal confidence levels and inflated test statistics that will have excessive type I error rates. It is essential to check for over-dispersion when fitting a generalized linear model or a generalized linear mixed model to ensure that inferences derived from the fitted model are accurate. Over-dispersion is an indication that the assumed model is incorrect and modifications are necessary.

The assumed model may be incorrect for various reasons. Generalized linear models consist of three components: the linear predictor, the distribution of the observed data, and the link. In addition, for a generalized linear mixed model the random effects have associated variance and covariance assumptions. While improper choice of any of these components can cause model misspecification, over-dispersion most frequently results either from assuming the wrong distribution for the observed data or from choosing an incorrect linear predictor. Examples of incorrectly specified linear predictors include those in which important predictors are missing from the model (e.g., leaving a quadratic term out of a model that is clearly not linear), predictors are expressed on the wrong scale [e.g., the response is not linear in X, but is linear on $\log(X)$], and random model effects are missing from the linear predictor (e.g., omitting the whole plot error term from a model for a split plot experiment).

Assuming the linear predictor is adequately defined, the most common cause of over-dispersion with count data is assuming the wrong distribution for the observed data. Count data are often treated as if they are Poisson distributed. Although the Poisson distribution has a long history of being taught as the distribution for count data in probability and statistics courses, there is an accumulating and increasingly persuasive body of evidence that the Poisson is not the distribution of choice for biological count data. There are compelling theoretical reasons as well. The Poisson distribution assumes that the events being counted occur in an independent and identically distributed fashion in time or space. For example, counts of weeds having a Poisson distribution would require the weeds to be equally likely to grow anywhere in a field. In contrast, most biological entities of interest to agronomists aggregate, or cluster, in some way. If the weeds are clustered, they are not distributed at random and, hence, cannot be considered as observations from a Poisson distribution. Observations in such aggregations, or groups, tend to be more alike than are observations from different groups. Therefore, the observations cannot be identically distributed. Distributions such as the negative binomial allow for this type of aggregation and tend to be better choices for biological count data.

The next two sections consider methods for recognizing and correcting for over-dispersion. Section 5.6 illustrates over-dispersion resulting from incorrectly specifying the distribution of the observed data. Over-dispersion in Section 5.7 results from an incorrectly specified linear predictor.

5.6 OVER-DISPERSION FROM AN INCORRECTLY SPECIFIED DISTRIBUTION

The examples in this section deal with over-dispersion resulting from an incorrect distributional assumption for the response. Since there are many probability distributions that describe processes that generate count data, it is often possible to resolve the over-dispersion problem by choosing an alternative distribution for the response. The first example presents some approaches to recognizing the existence of over-dispersion in the data. The other two examples consider different strategies for addressing its effect.

EXAMPLE 5.9

The data for this example are from an experiment to compare different cultural practices and seed mixes in attempting to restore damaged areas in the Nebraska Sand Hills (Stroup and Stubbendieck, 1983). Four fields were each divided into seven plots, and each cultural practice was assigned at random to one plot per field. Each cultural practice plot was subdivided into four smaller plots, and the seed mixes were randomly assigned to the smaller plots within each larger plot. Thus, the experiment was conducted as a split plot with a 7×4 factorial treatment structure with the whole plots in blocks. The response variable was the count of a plant species of interest.

Cultural practices (*practice*) and seed mixes (*mix*) were assumed to be fixed effects, and blocks (*block*) and whole plot error were random effects. Both random effects were assumed to be independent, normally distributed with mean zero and variances σ_B^2 and σ_W^2, respectively. The conditional mean of the plant species count, Y, is given by

$$\eta_{ijk} = g(E[Y_{ijk} \mid B_k, w_{ik}]) = \beta_0 + P_i + M_j + PM_{ij} + B_k + w_{ik}, \ i = 1, \ldots, 7; j = 1, 2, 3, 4; k = 1, 2, 3, 4$$

where β_0 is the overall mean, P_i is the effect of the ith cultural practice, M_j is the effect of the jth seed mix, PM_{ij} is the cultural practice $\times$ seed mix interaction effect, B_k is the kth block effect, and w_{ik} is the whole plot error associated with the ith cultural practice in the kth block.

The probability distribution of the observed count, Y, conditional on the random effects should follow from the assumptions about the processes that produce the plants in the plots. The linear predictor and random effects in the model follow from the treatment structure and experiment design and remain the same regardless of the distribution of the observed counts. The only questions to be resolved in this example are: What distribution best describes these data? And how does one decide?

FIG. 5-48. GLIMMIX statements to fit the initial Poisson model using the Laplace method for Example 5.9.

```
proc glimmix data=od method=laplace plots=studentpanel;
  class practice block mix;
  model count = practice mix practice*mix / dist=Poisson link=log;
  random intercept practice / subject=block;
  output out=new1 pred=pred stderr=sepred resid=resid student=student;
run;
```

Initially we will assume that the observed counts have a Poisson distribution with parameter $\lambda_{ijk} = E[Y_{ijk} \mid B_k, w_{ik}]$. The link function for the Poisson distribution is the natural logarithm; i.e., $\eta_{ijk} = g(\lambda_{ijk}) = g(E[Y_{ijk} \mid B_k, w_{ik}]) = \log(E[Y_{ijk} \mid B_k, w_{ik}])$.

The GLIMMIX statements for the initial fit of the Poisson model are shown in Fig. 5–48. Either the Laplace or quadrature methods must be specified in the *method* option on the *model* statement so that crucial over-dispersion diagnostic statistics will be calculated correctly. Both methods require the *subject* option form for all *random* statements.

The basic model and fitting information are presented in Fig. 5–49. As in the split plot design in Example 5.6, the G-side covariance parameters are the block and whole plot error variances and there are no R-side covariance parameters because there is no scale parameter associated with the Poisson distribution or equivalently, $\phi = 1$.

The goodness of fit statistics for the conditional distribution are shown in Fig. 5–50. Recall from Example 5.6 that the adequacy of the Poisson assumption can be assessed by the Pearson chi-square statistic divided by its degrees of freedom. If the Poisson fits perfectly, the Pearson chi-square/df would equal one. Values substantially greater than one suggest over-dispersion. This is not a formal hypothesis test, so there is no formal criterion such as a *p*-value for assessing the fit. As a very general guideline based on experience, Pearson chi-square/df values greater than two are potential indicators of over-dispersion. Certainly a value of 4.50 is a strong indication of over-dispersion.

The fit statistics in Fig. 5–50 depend on the use of the Laplace or quadrature methods. If either of these methods had not been specified on the GLIMMIX statement in Fig. 5–48, the default pseudo-likelihood procedure would have been used to fit the model. Since pseudo-likelihood is based on a Taylor series approximation to the conditional likelihood and not explicitly on the conditional likelihood itself (Section 2.7), a goodness of fit statistic such as the Pearson chi-square that is specifically applicable to the conditional distribution cannot be computed. Instead, the pseudo-likelihood approach calculates a generalized chi-square statistic that measures the combined fit of the conditional distribution of the counts and the random effects of the blocks and whole plot error. Because it is not specific to only the conditional distribution, it does not provide a clear cut diagnostic to assess the fit of the Poisson distribution to the counts.

To illustrate this point, the initial model was refit using the default pseudo-likelihood by removing the *method* option specified on the GLIMMIX statement in

FIG. 5-49. GLIMMIX output containing the basic model and fitting information using the Laplace method for the initial Poisson model for Example 5.9.

Model Information

Data Set	WORK.OD
Response Variable	Count
Response Distribution	Poisson
Link Function	Log
Variance Function	Default
Variance Matrix Blocked By	block
Estimation Technique	Maximum Likelihood
Likelihood Approximation	Laplace
Degrees of Freedom Method	Containment

Class Level Information

Class	Levels	Values
practice	7	1 2 3 4 5 6 7
block	4	1 2 3 4
mix	4	1 2 3 4

Number of Observations Read	112
Number of Observations Used	112

Dimensions

G-side Cov. Parameters	2
Columns in X	40
Columns in Z per Subject	8
Subjects (Blocks in V)	4
Max Obs per Subject	28

FIG. 5-50. GLIMMIX output containing the conditional fit statistics using the Laplace method for the initial Poisson model for Example 5.9.

Fit Statistics for Conditional Distribution

-2 log L(count | r. effects)	1053.96
Pearson Chi-Square	504.43
Pearson Chi-Square / DF	4.50

Fig. 5–48. The remaining statements were unchanged. The resulting goodness of fit statistics are shown in Fig. 5–51. The values of the chi-square statistics are larger than the corresponding values in Fig. 5–50, reflecting the additional effects of the blocks and whole plot error. These numerical values cannot be broken down into components reflecting the adequacy of the Poisson fit and the additional effect of the random effects.

In addition to the conditional fit statistics, another diagnostic that can help visualize over-dispersion in the Poisson is a graph of the variance versus the mean

FIG. 5-51. GLIMMIX output containing the conditional fit statistics using the default pseudo-likelihood method for the initial Poisson model in Example 5.9.

Fit Statistics	
-2 Res Log Pseudo-Likelihood	548.62
Generalized Chi-Square	525.55
Gener. Chi-Square / DF	6.26

FIG. 5-52. Graph of the sample mean versus the sample variance for each cultural practice–seed mix combination from the initial Poisson model for Example 5.9. The dashed line represents the mean equal to the variance as would be expected for the Poisson distribution.

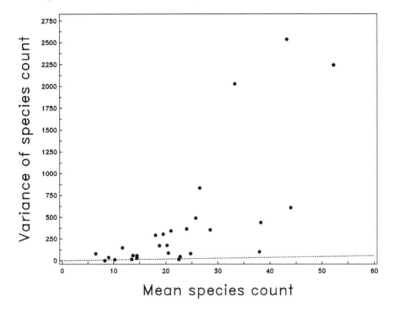

Mean species count

count for each cultural practice–seed mix combination. The plot is shown in Fig. 5–52. Under the Poisson assumption, the mean and variance are equal (represented by the dashed line near the bottom of the graph). The graph clearly shows variances much greater than the corresponding means, providing strong visual evidence of considerable over-dispersion and a clear indication that the Poisson assumption is not reasonable for these data. ∎

When there is a strong indication of over-dispersion, as there is with the data in Example 5.9, some action must be taken to avoid the undesirable consequences described previously. Two strategies appear prominently in the generalized linear mixed model literature; namely, adjust the standard errors and test statistics by including an adjustment for over-dispersion in the model, or assume a different probability distribution for the counts that more realistically approximates the process by which they arise. The former is a technique borrowed from generalized linear models having only fixed effects. While it seems reasonable, experience has led to a consensus among GLMM users that the adjustment by the over-dispersion parameter approach is not a viable method for GLMMs. Because it is so prevalent in the literature, it is presented in Example 5.10 as an example of "what not to do" so that readers will be able to recognize it. The second approach of assuming a different distribution will be illustrated in Example 5.11. It is a viable and recommended strategy.

EXAMPLE 5.10

This example is a continuation of Example 5.9 in which an over-dispersion parameter will be added to the Poisson model. This is the "what not to do" illustration. Its basis in generalized linear model theory is as follows. For models with distributions for which the scale parameter ϕ is equal to one, the addition of a multiplicative over-dispersion parameter is equivalent to removing the restriction on the scale parameter. For distributions that have an unconstrained scale parameter, the over-dispersion parameter adds a multiplicative scalar to the variance function. Inclusion of an over-dispersion parameter does not affect the parameter estimates. However, it does change the estimated covariance matrix of the estimates by a scalar multiple. Tables 2–1 and 2–2 list information on the scale parameter for distributions used in generalized linear mixed models.

Since the scale parameter in the Poisson is fixed at one, adding the over-dispersion parameter is equivalent to removing the constraint from the scale parameter. Hence, the conditional variance of the count Y_{ijk} becomes

$$\text{var}(Y_{ijk} \mid B_{k}, w_{ik}) = \phi \lambda_{ijk}$$

where ϕ is the scale or over-dispersion parameter. The over-dispersion parameter measures the magnitude of over-dispersion. Its estimate is analogous to the mean square error in a normal theory analysis of variance.

The scale parameter approach is the classical fix for over-dispersion in Poisson regression and analysis of variance models. It is described in detail in generalized linear model textbooks such as McCullagh and Nelder (1989). However, this method has two major drawbacks. First, there is no probability distribution that has the Poisson form with a mean of λ_{ijk} and a variance of $\phi \lambda_{ijk}$. The introduction of the over-dispersion parameter forces one to assume a quasi-likelihood, a random variable whose structure resembles a probability distribution but in fact has no known distribution. This means that there is no mechanism by which random variables of this type could arise. The second drawback of such models is that simulation work has cast doubt on the ability of the scale parameter approach in the Poisson to adequately control type I error and provide accurate coverage for confidence intervals. For these reasons, we recommend finding an alternative distribution if at all possible.

The GLIMMIX statements to fit the Poisson distribution with over-dispersion are shown in Fig. 5–53. The second *random* statement that contains the keyword _residual_ causes the scale parameter to be estimated and used to adjust standard

FIG. 5-53. GLIMMIX statements to fit the Poisson model with over-dispersion using quasi-likelihood for Example 5.10.

```
proc glimmix data=od  plots=studentpanel;
    class practice block mix;
    model count = practice mix practice*mix / dist=Poisson link=log;
    random intercept practice / subject=block;
    random _residual_ ;
    output out=new3 pred=pred stderr=sepred resid=resid student=student;
run;
```

FIG. 5-54. GLIMMIX output containing the *Dimensions* section for the Poisson model with over-dispersion using quasi-likelihood for Example 5.10.

Dimensions	
G-side Cov. Parameters	2
R-side Cov. Parameters	1
Columns in X	40
Columns in Z per Subject	8
Subjects (Blocks in V)	4
Max Obs per Subject	28

FIG. 5-55. GLIMMIX output containing the covariance and over-dispersion (scale) parameter estimates in the Poisson model with over-dispersion using quasi-likelihood for Example 5.10.

Covariance Parameter Estimates			
Cov Parm	Subject	Estimate	Standard Error
Intercept	block	0.01706	0.05573
practice	block	0.2334	0.1096
Residual (VC)		7.5955	1.3433

errors and test statistics. Note that this program cannot be run using the Laplace or quadrature methods; it must be run using the default method only.

The *Dimensions* section of the GLIMMIX output is shown in Fig. 5–54. As before, the G-side covariance parameters are the block and whole plot error variances. The R-side covariance parameter is the scale or over-dispersion parameter. The estimates are presented in Fig. 5–55. The scale parameter estimate $\hat{\phi}$ of 7.5955 is used to adjust all fixed effects standard errors and test statistics.

To illustrate the impact of the over-dispersion parameter, the tests of the fixed effects are shown in Fig. 5–56. In contrast, the tests for the fixed effects from the fit using the default pseudo-likelihood without over-dispersion in Example 5.9 are

FIG. 5-56. GLIMMIX output containing the tests for the fixed effects in the Poisson model with over-dispersion using quasi-likelihood for Example 5.10.

Type III Tests of Fixed Effects

Effect	Num DF	Den DF	F Value	Pr > F
practice	6	18	1.63	0.1971
mix	3	63	3.02	0.0364
practice*mix	18	63	1.32	0.2043

FIG. 5-57. GLIMMIX output containing the tests of the fixed effects using the default pseudo-likelihood method for the initial Poisson model for Example 5.9.

Type III Tests of Fixed Effects

Effect	Num DF	Den DF	F Value	Pr > F
practice	6	18	1.75	0.1675
mix	3	63	22.91	<0.0001
practice*mix	18	63	10.06	<0.0001

shown in Fig. 5–57. The impact on the split plot terms, the seed mix main effect and cultural practice × seed mix interaction, is especially striking. Their F values from the fit without over-dispersion are divided by approximately 7.6, the scale parameter estimate, to obtain the F values in Fig. 5–56. While including an over-dispersion parameter changes the standard errors, F values, and p-values, it does not do so correctly, as will be shown, and, hence, should not be used. ∎

EXAMPLE 5.11

This example is a continuation of Example 5.9 in which the negative binomial distribution replaces the Poisson distribution as the conditional distribution of the response. This is an example of a viable generalized linear mixed model strategy for dealing with over-dispersion. Its basis in generalized linear mixed model methodology is as follows.

The leading candidate for an alternative to the Poisson is the negative binomial distribution (Section 2.3). Unlike the Poisson with a scale parameter, the negative binomial is an actual probability distribution, which means that there is a plausible mechanism in terms of probability theory by which counts showing behavior that follows the negative binomial could arise. Moreover, the mathematical derivation of the negative binomial assumes an aggregation process rather than a completely independent process as in the Poisson. Aggregation is often more realistic for biological count processes. Like the Poisson, the negative binomial has a mean of λ, but the variance is $\lambda + \phi\lambda^2$, where ϕ is a scale parameter that is part of the definition of the distribution. The negative binomial scale parameter should not be confused with the over-dispersion parameter added to the Poisson model in Example 5.10. Depending on the value of ϕ, the variance is free to take on any value greater than λ.

The Laplace and quadrature methods are preferred for fitting the negative binomial. There are two reasons for using one of these methods. First, they allow the Pearson chi-square statistic to be calculated to assess the negative binomial's appropriateness as the conditional distribution of the counts. Second, the Laplace and quadrature methods fit the actual likelihood of the negative binomial by the most direct available method, resulting in a more accurate fit of the model.

The GLIMMIX statements used to fit the negative binomial distribution are shown in Fig. 5–58, and the basic model and fitting information is presented in Fig. 5–59. The R-side covariance parameter is the scale parameter ϕ that is part of the definition of the negative binomial distribution. It is not an over-dispersion parameter as in Example 5.10.

The fit statistics for the conditional distribution of the response are shown in Fig. 5–60. The Pearson chi-square/df value of 0.71 indicates that the negative binomial provides a much improved fit of the data compared to the Poisson in Example 5.9.

The covariance parameter estimates are shown in Fig. 5–61. The estimated scale parameter is 0.3459, so that the estimated conditional variance of the count is $\hat{\lambda}_{ijk} + 0.35\hat{\lambda}_{ijk}^2$, where $\hat{\lambda}_{ijk}$ is the conditional mean on the data (count) scale. The tests for the fixed effects are presented in Fig. 5–61. These F values are more in line with the over-dispersed adjusted Poisson fixed effects tests shown in Fig. 5–56. However, there are some important differences, but, as discussed in Example 5.10, the use of an over-dispersed Poisson is not appropriate here, and we will not explore these differences. Unlike the Poisson, in the negative binomial there is statistically significant evidence of a cultural practice main effect and less convincing evidence of a main effect of seed mix ($p = 0.1055$). ∎

There are two other viable approaches for these data that are not shown in this context but are illustrated elsewhere in this chapter. These approaches can be summarized as follows.

- Following the skeleton ANOVA approach introduced in Example 2.10 and applied to the conditional GLMM in Example 5.1, fit a model assuming a Poisson distribution and add "the last line of the ANOVA"

FIG. 5–58. GLIMMIX statements to fit the negative binomial distribution using the Laplace method for Example 5.11.

```
proc glimmix data=od  method=laplace  plots=studentpanel;
class practice block mix;
model count = practice mix practice*mix / dist=NegBin link=log;
random intercept practice / subject=block;
output out=new4  pred=pred stderr=sepred resid=resid student=student;
run;
```

FIG. 5-59. GLIMMIX output containing the basic model and fitting information using the Laplace method for the negative binomial model for Example 5.11.

Model Information	
Data Set	WORK.OD
Response Variable	count
Response Distribution	Negative Binomial
Link Function	Log
Variance Function	Default
Variance Matrix Blocked By	block
Estimation Technique	Maximum Likelihood
Likelihood Approximation	Laplace
Degrees of Freedom Method	Containment

Class Level Information		
Class	Levels	Values
practice	7	1 2 3 4 5 6 7
block	4	1 2 3 4
mix	4	1 2 3 4

Number of Observations Read	112
Number of Observations Used	112

Dimensions	
G-side Cov. Parameters	2
R-side Cov. Parameters	1
Columns in X	40
Columns in Z per Subject	8
Subjects (Blocks in V)	4
Max Obs per Subject	28

FIG. 5-60. GLIMMIX output containing the fit statistics for the negative binomial model in Example 5.11 using the Laplace method.

Fit Statistics for Conditional Distribution	
$-2 \log L(count \mid r.\ effects)$	838.51
Pearson Chi-Square	79.35
Pearson Chi-Square / DF	0.71

to the linear predictor. Note that this term is a random effect. Recall in Example 5.1 that failing to include the block × variety interaction in the model introduced a form of over-dispersion resulting in a severely biased F value and standard errors characteristic of over-dispersion. Adding the block × variety interaction random effect to the linear predictor solved the problem in that example. The analogous term for the model in Example 5.9 is the cultural practice × seed mix × block interaction, PMB_{ijk}.

FIG. 5-61. GLIMMIX output containing the covariance parameter estimates and tests for the fixed effects for the negative binomial model in Example 5.11 using the Laplace method.

Covariance Parameter Estimates

Cov Parm	Subject	Estimate	Standard Error
Intercept	block	0.002390	0.02764
practice	block	0.1222	0.07104
Scale		0.3459	0.06877

Type III Tests of Fixed Effects

Effect	Num DF	Den DF	F Value	Pr > F
practice	6	18	2.70	0.0473
mix	3	63	2.13	0.1055
practice*mix	18	63	1.18	0.3019

- The negative binomial and the above Poisson model with PMB_{ijk} included are both conditional models. Recall from Example 5.2 that an alternative was the marginal model. If inference based on the marginal model was deemed appropriate for Example 5.9, the random whole plot effect could be replaced with a compound symmetry working covariance. Compound symmetry does include a scale parameter that can be interpreted as an over-dispersion parameter. There are two caveats for this approach. First, use it only if marginal model based inference is deemed appropriate; i.e., do not think of it as an over-dispersion tactic. Second, use it in conjunction with the Poisson, where it forms a Poisson-based quasi-likelihood that is reasonably well understood.

The examples in this and the previous section provide a general strategy for analyzing generalized linear mixed models with potentially over-dispersed count data. First, check for over-dispersion using the Laplace or quadrature methods. If there is evidence that over-dispersion is present in the data, then use an alternative distribution if at all possible. Adding a "last line of the ANOVA" G-side random effect to the Poisson model, as was shown in the binomial model in Example 5.1, can also be a viable strategy. Adding an over-dispersion parameter to the original distribution is strongly discouraged. The marginal GEE compound symmetry Poisson model would be the alternative of choice if inference based on the marginal distribution was deemed to be best suited to addressing study objectives.

5.7 OVER-DISPERSION FROM AN INCORRECT LINEAR PREDICTOR

Over-dispersion because of an incorrect linear predictor probably is more common in regression than analysis of variance, especially in those problems where subject matter considerations do not mandate the functional form of the regression. Incorrect linear predictors in analysis of variance situations are more likely to occur because of unrealistic assumptions concerning the process generating the data. This latter situation will be illustrated in the examples in this section.

EXAMPLE 5.12

Suppose that 10 technicians are given the task of evaluating the germination rate π of a given lot of corn seed. Each technician starts with $N = 100$ seeds and uses the same germination protocol. After the ith technician completes the protocol, she computes the sample proportion $p_i = Y_i/100$, where Y_i is the number of seeds in her sample of 100 that germinate.

Consider two possible scenarios for modeling this experiment.

Scenario 1: Assume that

- the sampling of the seed samples from the lot is completely random,
- seeds germinate independently; i.e., the outcome for one seed does not affect whether or not any other seed germinates,
- the probability that a seed germinates remains constant for all seeds in a technician's sample,
- technicians have no effect on the outcome of germination; i.e., the probability that a seed germinates is not dependent on the technician who is evaluating the sample.

Model 1: It follows from the assumptions listed above that the Y_i have binomial distributions. These assumptions establish a distribution for a generalized linear model. Because all technicians are estimating the same germination rate, the linear predictor is $\text{logit}(\pi) = \log[\pi/(1-\pi)]$. The model used to fit these data could be simply stated as Y_i is binomial($N = 100$, π) with $E[p_i] = \pi$. Fitting this model reduces to computing $p = (\sum_{i=1}^{10} Y_i)/1000$.

Scenario 2: Assume that

- the sampling of the seed samples from the lot is completely random,
- seeds germinate independently; i.e., the outcome for one seed does not affect whether or not any other seed germinates,
- the probability that a seed germinates remains constant for all seeds in a technician's sample,
- the technician does have an effect on the outcome of germination; i.e., the probability that a seed germinates is dependent on the technician who is evaluating the sample,
- the technician effects are independent and normally distributed with mean zero and variance σ_T^2.

Model 2: It follows from the assumptions that the distribution of the data conditional on the ith technician, $Y_i \mid T_i$, is binomial(100, π_i) where the π_i depend on the technician and vary randomly. These assumptions establish a distribution for a generalized linear mixed model with the random effect attributable to the technicians. For this model the linear predictor can be written as $\text{logit}(\pi_i) = \log[\pi_i/(1 - \pi_i)] = \beta_0 + T_i$, where T_i is the random effect of the ith technician. The model used to fit these data could be simply stated as $Y_i \mid T_i$ is binomial(100, π_i) with $T_i \sim N(0, \sigma_T^2)$. ∎

Both scenarios in Example 5.12 are plausible for describing how counts could be generated from a germination study, but Scenario 1 does not include any technician effects and should be used only if one is willing to make the strong assumption that technicians have no impact on the experiment. As a consequence, there are noticeable differences in the models that are used to explain the scenarios. Model 1 is a generalized linear model with no technician effect, and Model 2 is a generalized linear mixed model that accounts for technician effects.

Let $\hat{p}_{si}$ denote the predicted value of the sample proportion for the ith technician under the sth scenario ($s = 1, 2$) and corresponding model, and let $\hat{\sigma}_{si}$ denote its estimated standard error. Let $r_i = (p_i - \hat{p}_{si})/\hat{\sigma}_{si}$ be the residual for the fitted sample proportion for the ith technician under the sth scenario. The statistic Σr_i^2 is Pearson's chi-square goodness of fit statistic. A fitted model is deemed a "good fit" when the value of the statistic is close to the number of residuals. This follows from the fact that when the r_i are approximately normal, $E[r_i^2]$ is approximately one, and hence, the mean of the sum is approximately the number of residuals. When the model is correctly specified, the normality assumption for the r_i is not required for the result to hold.

There is a less obvious difference between the models that has to do with the difference between model parameters that can be used to describe them. Model 2 is posed in terms of a conditional distribution and includes a specification for the distribution of the random effects, whereas Model 1 is strictly unconditional. Under Model 1, the unconditional expected value or mean of a sample proportion p_i is π where π, the probability of a seed being counted as germinated, does not vary among technicians.

Model 2 describes a two stage process. The first stage of the process involves randomly selecting a probability of success π_i for the ith technician from a probability distribution whose mean π drives the process. The second stage of the process involves generating a count Y_i for the ith technician from the binomial distribution based on the probability of success π_i.

Note that the process for Model 2 is a one treatment version of the ten block, two variety example begun in Section 2.9 and revisited in Section 5.3 to illustrate the difference between conditional and marginal modeling in generalized linear mixed models. Model 2 is the conditional logit generalized linear mixed model for this process. Later we will introduce a marginal model, Model 3, which accounts for technician variance on the R-side similar to the GEE shown in Example 5.2. Models 1 and 3 will estimate the mean of the marginal distribution as illustrated in Fig. 5–9, whereas Model 2 will estimate the conditional mean as defined in Section 5.3. To distinguish between these estimators, denote the estimated marginal mean obtained from Models 1 and 3 by $\hat{p}_{1\cdot}$ and denote the conditional mean obtained from Model 2 by $\hat{p}_{2\cdot}$. Recall from Example 5.1 that the expected value of the marginal estimate $\hat{p}_{1\cdot}$ is not equal to π, the probability in the process described by Model 2, whereas expected value of the estimate $\hat{p}_{2\bullet}$ based on the conditional model does equal π.

To summarize, the marginal (unconditional) mean of the sample proportion is not the same as the probability identified with the process for Model 2. This is important to distinguish because parameter estimates reported from fitting Model 1 will have a different interpretation from those reported when fitting Model 2.

Finally we describe the difference between the two scenarios by comparing the expected variance of the total number of seeds that germinate, ΣY, based on the scenarios' corresponding assumptions. For sake of generality, let K denote the number of technicians used in the germination study and, as before, let N be the number of seeds evaluated by each technician. It follows from the assumptions of Scenario 1 that $\Sigma Y_i \sim$ binomial(KN, π) so that Var(ΣY_i) = $KN\pi(1 - \pi)$, where π = E[$\Sigma Y_i/(KN)$].

Based on the assumptions of Scenario 2, Faraway (2006) shows that Var(ΣY_i) is inflated by the technician effect. The inflated variance will be too large to be consistent with a binomial distribution. This result demonstrates an important point. The inflated variance resulting from the Scenario 2 assumptions is directly attributable to the random effects of technicians. Thus, including random effects in a generalized linear mixed model is a way to account for over-dispersion.

EXAMPLE 5.13

Suppose that the data from the germination study in Example 5.12 are as given in Table 5–1. The GLIMMIX statements to fit Models 1 and 2 under their respective scenarios are shown in Fig. 5–62. In the *model* statement for both scenarios, the lack of terms after the equal sign indicates that only a single intercept parameter will be fit. The *covtest* option *zerog* in Model 2 tests for the G-side covariance parameter (technician variance) equal to zero.

The *Dimensions* sections of the GLIMMIX outputs for Models 1 and 2 are shown in Fig. 5–63 and 5–64, respectively. The column in the X matrix represents the intercept or overall mean. Model 1 has neither G-side nor R-side covariance parameters, as indicated by their absence from the section. For Model 2, the G-side parameter is the technician variance.

The primary difference between the fit statistics for Models 1 and 2 results from the fact that Model 2 includes technician as a random effect but Model 1 does not. The values of the Pearson chi-square statistics differ substantially. For Model 1, the value of the Pearson chi-square/df is 8.08 (Fig. 5–65), indicating over-dispersion. Fitting the same

TABLE 5–1. Number of seeds germinated by technician for Example 5.13.

Technician	Number of germinated seeds	Number of seeds tested
1	81	100
2	66	100
3	94	100
4	74	100
5	96	100
6	93	100
7	80	100
8	86	100
9	88	100
10	99	100

FIG. 5–62. GLIMMIX statements to fit the models associated with Scenarios 1 and 2 for Example 5.13.

```
title2 'MODEL 1 FITTED UNDER SCENARIO 1';
proc glimmix data=germ method=laplace;
  class tech;
  model germ/total =  ;
  estimate 'Mean proportion' intercept 1 / ilink;
run;


title2 'MODEL 2 FITTED UNDER SCENARIO 2';
proc glimmix data=germ method=laplace;
  class tech;
  model germ/total =  ;
  random intercept / subject=tech;
  estimate 'Mean proportion' intercept 1 / ilink;
  covtest 'Technician effect = 0' zerog;
run;
```

FIG. 5–63. GLIMMIX output containing the *Dimensions* section from the fit of Model 1 for Example 5.13.

Dimensions	
Columns in X	1
Columns in Z	0
Subjects (Blocks in V)	1
Max Obs per Subject	10

FIG. 5–64. GLIMMIX output containing the *Dimensions* section from the fit of Model 2 for Example 5.13.

Dimensions	
G-side Cov. Parameters	1
Columns in X	1
Columns in Z per Subject	1
Subjects (Blocks in V)	10
Max Obs per Subject	1

data but accounting for technicians as a random effect in Model 2 seems to ameliorate the over-dispersion problem since the Pearson chi-square/df value of 0.18 is much less than one (Fig. 5–66). In Fig. 5–67 the variance associated with the technician effect in Model 2 is estimated to be 0.8364. The highly significant p-value for testing $\sigma_T^2 = 0$ further substantiates the impact that technicians have on explaining the data. Scenario 2 is certainly a more plausible explanation for the observed data than Scenario 1 based on the goodness of fit statistics and the highly significant technician random effect.

The estimated mean proportions in Fig. 5–68 and 5–69 in the column labeled *Mean* are different. The reason for this difference is important to understand. Under Model 1 it is an estimate of the marginal or unconditional mean. This represents an estimate for the expected proportion over the entire population of technicians. The estimate reported for Model 2 is an estimate for the expected proportion for the average technician, $E[p_i \mid T_i = 0]$, which is an estimate for the probability π as

FIG. 5-65. GLIMMIX output containing the fit statistics for Model 1 for Example 5.13.

Fit Statistics

-2 Log Likelihood	125.36
AIC (smaller is better)	127.36
AICC (smaller is better)	127.86
BIC (smaller is better)	127.66
CAIC (smaller is better)	128.66
HQIC (smaller is better)	127.03
Pearson Chi-Square	80.79
Pearson Chi-Square / DF	8.08

FIG. 5-66. GLIMMIX output containing the fit statistics for Model 2 for Example 5.13.

Fit Statistics for Conditional Distribution

-2 log L(germ | r. effects)	42.23
Pearson Chi-Square	1.78
Pearson Chi-Square / DF	0.18

FIG. 5-67. GLIMMIX output containing the estimated technician variance and test for the variance equal to zero for Model 2 for Example 5.13.

Covariance Parameter Estimates

Cov Parm	Subject	Estimate	Standard Error
Intercept	tech	0.8364	0.4685

Tests of Covariance Parameters Based on the Likelihood

Label	DF	−2 Log Like	ChiSq	Pr > ChiSq	Note
Technician effect = 0	1	125.36	52.59	<0.0001	MI

MI: P-value based on a mixture of chi-squares.

defined in the Model 2 process. As discussed in Sections 2.10 and 5.3 and illustrated in Fig. 5–9, the marginal and conditional estimates differ as expected.

Which of the estimated means should we report? If Model 1 did not show evidence of over-dispersion—if it did indeed fit the data—the question would be moot. There is no estimate of $E[p_i \mid T_i = 0]$ for Model 1 because there is no technician random effect specified. However, if Model 2 is used for fitting the data and if the process giving rise to the data described by Model 2 fits the data, then the question would be worth considering. Should we report the conditional estimate from Model 2 or replace it by a marginal estimate that also accounts for variability among technicians? ■

Example 5.13 poses a more general problem. The fit of the data to Model 1 reports an estimate for a marginal mean but suffers from over-dispersion. The consequence of the poor fit is that the estimate's standard error is underestimated. Inferences on the marginal mean using estimates from Model 1 would be severely inaccurate. The standard error for the estimate for Model 2 is larger, at least three

FIG. 5–68. GLIMMIX output containing the estimated mean proportion for Model 1 for Example 5.13.

Label	Estimate	Standard Error	DF	t Value	Pr > |t|	Mean	Standard Error Mean
Estimates							
Mean proportion	1.7906	0.09033	9	19.82	<0.0001	0.8570	0.01107

FIG. 5–69. GLIMMIX output containing the estimated mean proportion for Model 2 in Example 5.13.

Label	Estimate	Standard Error	DF	t Value	Pr > |t|	Mean	Standard Error Mean
Estimates							
Mean proportion	2.0580	0.3134	9	6.57	0.0001	0.8868	0.03147

times larger in the example. Model 2 fits the data much better than Model 1 by accounting for substantial technician effects, but the output reports an estimate for the expected proportion for an average technician. We need to decide if our objectives are best addressed by (i) estimating a marginal germination probability averaged over all technicians in the population (assuming that the sample of technicians in the experiment accurately represents the target population) or (ii) estimating a conditional germination probability for an average technician.

If the second option meets our needs, we are done; we use the conditional estimate from Model 2. If our objectives call for the first option, then we need to use a marginal model, moving the technician variance to the R-side to account for over-dispersion and removing technician effects from the linear predictor. For this, we use the generalized estimating equation (GEE) approach introduced in Section 5.2.

EXAMPLE 5.14

A simple GEE approach uses the marginal linear predictor from Model 1 in Example 5.13 but adds an over-dispersion or scale parameter ϕ to account for over-dispersion and, as a result, adjusts the standard error appropriately. The correction uses this scale estimate to adjust the estimated standard error of the marginal mean obtained from the fit of Model 1 by a factor of $\sqrt{\hat{\phi}}$. Let Model 3 denote the GEE approach to Scenario 1.

The GLIMMIX statements to fit Model 3 are shown in Fig. 5–70. The *method* option on the PROC GLIMMIX statement indicates that a pseudo-likelihood procedure will be used to obtain the estimates. As in Example 5.10, the *random* statement adds the over-dispersion or scale parameter to the model.

The *Dimensions* section of the GLIMMIX output is shown in Fig. 5–71. The R-side covariance parameter is the scale parameter. The model contains no random effects, as indicated by the absence of G-side parameters and zero columns in the **Z** matrix.

The fit statistics are shown in Fig. 5–72 and the estimated scale parameter in Fig. 5–73. The estimated scale parameter $\hat{\phi}$ = 8.0791 in Fig. 5–73 is identical to the Pearson chi-square/df value in Fig. 5–72. The value of the estimated proportion of germinated seed, 0.8570 from Fig. 5–74, is the same as reported for Model 1 in Fig. 5–68, but its standard error has increased from 0.01107 to 0.03147, an increase by a factor of $\sqrt{\hat{\phi}} = \sqrt{8.079}$. Note that the corrected standard error using Model 3 for Scenario 1 now matches the estimated standard error using Model 2 for Scenario 2, but the estimated proportion is not conditional on technician. The appropriate confidence limits for the true marginal proportion of seeds that germinate are determined by taking the inverse link of the confidence limits determined from the estimated logit and its corrected standard error from Model 1, namely 0.09033 $\sqrt{\hat{\phi}}$ = 0.2568 in Fig. 5–74. ∎

The examples in this section illustrated that over-dispersion may result from misspecification of the linear predictor. Omitting random effects will lead to over-dispersion when the variation due to these effects is substantial. Statistics needed for inference on marginal means can be computed by incorporating the estimated over-dispersion or scale parameter into the calculations.

5.8 EXPERIMENTS INVOLVING REPEATED MEASURES

Repeated measures analysis for linear mixed models with normally distributed data was discussed in Section 4.6. In this section, we consider repeated measures with non-normal data. Data of this type arise from the same kind of experimental designs as normally distributed repeated measures data. The only difference is that the response conditional on the random effects is non-normal.

The examples in this section use the binomial distribution with the logit link but are equally applicable to the probit link or the other links identified in Table

FIG. 5–70. GLIMMIX statements to fit Model 3 to Scenario 1 for Example 5.14.

```
proc glimmix data=germ method=mmpl;
    class tech;
    model germ/total = ;
    estimate 'Mean proportion' intercept 1 / ilink cl;
    random _residual_ / subject=tech;
run;
```

FIG. 5–71. GLIMMIX output containing the *Dimensions* section from the fit of Model 3 to Scenario 1 for Example 5.14.

Dimensions	
R-side Cov. Parameters	1
Columns in X	1
Columns in Z per Subject	0
Subjects (Blocks in V)	10
Max Obs per Subject	1

FIG. 5-72. GLIMMIX output containing the fit statistics from the fit of Model 3 to Scenario 1 for Example 5.14.

FIG. 5-73. GLIMMIX output containing the estimated covariance parameter from the fit of Model 3 to Scenario 1 for Example 5.14.

Fit Statistics

−2 Log Pseudo-Likelihood	24.21
Generalized Chi-Square	80.79
Gener. Chi-Square / DF	8.08

Covariance Parameter Estimates

Cov Parm	Estimate	Standard Error
Residual (VC)	8.0791	3.6131

FIG. 5-74. GLIMMIX output containing the estimated mean proportion from the fit of Model 3 ($\alpha = 0.05$) to Scenario 1 for Example 5.14.

Estimates

Label	Estimate	Standard Error	DF	t Value	Pr > |t|	Lower	Upper	Mean	Standard Error Mean	Lower Mean	Upper Mean
Mean proportion	1.7906	0.2568	9	6.97	<0.0001	1.2098	2.3714	0.8570	0.03147	0.7703	0.9146

3.1. In addition, the methods shown in this section may be used with any of the probability distributions described in Section 2.3 for generalized linear mixed models. Transition from the binomial to other one parameter distributions, for example the Poisson distribution, involves only replacing the assumed distribution. Distributions that also involve a scale parameter, such as the negative binomial or beta are essentially straightforward as long as the meaning of the scale parameter is preserved.

As with other generalized linear mixed model examples, there is a marginal model and a conditional model for repeated measures. These are also called the R-side and G-side approaches, respectively. The marginal or R-side model builds on the GEE approach described in Section 5.3. Modeling non-normal repeated measures using GEE is the better known of the two approaches in the statistics literature. This is partly because the model borrows directly from the normal distribution repeated measures analysis with virtually no modification and partly because this approach appeared in the statistics literature before conditional correlated error models. Readers who have used SAS PROC GENMOD for non-normal repeated measures data used GEE. The conditional approach, also called the "true" generalized linear mixed model (GLMM) approach for repeated measures, is not available with GENMOD. In SAS only GLIMMIX performs the conditional model analysis. In general, only generalized linear model software having the capability of explicitly including random effects can implement conditional models. As has been discussed in previous examples, the G-side and R-side approaches each have distinct advantages and disadvantages.

Recall from Section 4.6 that the repeated measures mixed model is similar to the model for a split plot experiment, except for the assumptions about the within-subject observations, that is, observations at different times on the same subject. In the split plot, these observations are assumed to be independent; in repeated measures, they are potentially correlated. The initial transition from normally distributed errors repeated measures to non-normal repeated measures mimics the transition from normal to non-normal split plot analyses; i.e., replace the response variable by the link and drop the residual (split plot or within subjects) error term.

EXAMPLE 5.15

The data for this example are from a larger study by C.S. Rothrock (used with permission) to determine the usefulness of fungicides applied at planting time to reduce the effect of seedling diseases in cotton on stand counts. The design was a randomized complete block with five blocks and four treatments, three fungicides and a no fungicide control. Each plot consisted of two rows with 200 seeds per row. Counts of healthy plants were taken at 12, 20, and 42 days after planting. The response was the number of healthy plants out of 400.

The fungicide treatments (*trt*) and sampling time (*time*) were treated as fixed effects, and blocks (*block*) and whole plot error were random effects. The stand count, Y, was assumed to have a binomial distribution with $n = 400$ and unknown probability of a healthy plant π. Using the logit link, the conditional mean of Y is given by

$$\eta_{ijk} = \log\left(\frac{\pi_{ijk}}{1 - \pi_{ijk}}\right) = \beta_0 + B_i + F_j + w_{ij} + T_k + FT_{jk}, i = 1,\ldots,5; j = 1,\ldots,4; k = 1,2,3$$

where β_0 is the overall mean, B_i is the *i*th block effect, F_j is the *j*th fungicide treatment effect, T_k is the *k*th time effect, FT_{jk} is the fungicide $\times$ time interaction effect, and w_{ij} is the whole plot error. In repeated measures terminology, the whole plot error is called the between subjects error and the residual variance (split plot error for normally distributed data) is called the within subjects error. ∎

In repeated measures with a binomial response, the residual or within subjects error variance depends strictly on π_{ijk}, more specifically, on $\pi_{ijk}(1 - \pi_{ijk})$. The absence of this error term in the logit model suggests the crux of the repeated measures generalized linear model specification. How does one account for within-subjects serial correlation? To answer this question, we first consider the "split plot in time."

Recall from Section 4.4 that the split plot in time has two equivalent forms for normally distributed response variables. The independent errors model (a GLMM) includes the random between subjects effects (whole plot errors) and assumes the within subjects effects are independent while the compound symmetry model (a GEE) does not include the random between subjects effects explicitly in the model but rather embeds them in a compound symmetry covariance structure for the within subjects effects. The logit model for the conditional mean of Y in Example 5.15 is the independent errors version of the model. In the independent errors model with three sampling times, the within subjects variances can be characterized by the covariance matrix

$$
\begin{bmatrix} \pi_{ij1}(1-\pi_{ij1}) & 0 & 0 \\ 0 & \pi_{ij2}(1-\pi_{ij2}) & 0 \\ 0 & 0 & \pi_{ij3}(1-\pi_{ij3}) \end{bmatrix} = \mathrm{diag}\left[\pi_{ijk}\left(1-\pi_{ijk}\right)\right]
$$

The compound symmetry version of the logit model in Example 5.15 is

$$
\eta_{ijk} = \log\left(\frac{\pi_{ijk}}{1-\pi_{ijk}}\right) = \beta_0 + B_i + F_j + T_k + FT_{jk}, i = 1,\ldots,5; j = 1,\ldots,4; k = 1,2,3
$$

and the within subjects covariance matrix is amended to

$$
\mathrm{diag}\left[\sqrt{\pi_{ijk}(1-\pi_{ijk})}\right]\begin{bmatrix} 1 & \rho & \rho \\ \rho & 1 & \rho \\ \rho & \rho & 1 \end{bmatrix}\mathrm{diag}\left[\sqrt{\pi_{ijk}(1-\pi_{ijk})}\right]
$$

This form of the model requires a working correlation matrix. The diagonal variance function matrix in the above expression is split into two halves, each composed of the square roots of the variance function. The working correlation matrix has a compound symmetry structure and is inserted between the two halves. As with previous examples of working correlation in this chapter, this does not define an actual covariance structure. It represents a quasi-likelihood that mimics the form such a structure would have if there were a real probability distribution associated with it. Working correlation is an important tool for repeated measures generalized linear models.

EXAMPLE 5.16

This example is a continuation of Example 5.15 in which the analysis of the two variations of the split plot in time will be compared.

The GLIMMIX statements to fit the independent errors and compound symmetry forms of the model are shown in Fig. 5–75. These program statements are identical to the statements used for normally distributed split plot models except that the response is replaced by the binomial *events/trials* form. For readers who have used PROC GENMOD for repeated measures note that the compound symmetry form is the GLIMMIX version of the statements you would use in GENMOD except compound symmetry would be specified with a *repeated* statement, and the block effect would have to be treated as a fixed effect because GENMOD has no provision for random effects.

Essentially the statements in Fig. 5–75 introduce the two main approaches to analyzing repeated measures generalized linear models with GLIMMIX.

FIG. 5-75. GLIMMIX statements to fit split plot in time logit models for Example 5.16.

```
title2 'SPLIT PLOT INDEPENDENT ERRORS FORM';
proc glimmix data=fung;
    class block trt time;
    model count/n = trt time trt*time / ddfm=kr;
    random intercept trt / subject=block;
    lsmeans trt*time / ilink;
run;

title2 'SPLIT PLOT COMPOUND SYMMETRY FORM';
proc glimmix data=fung;
    class block trt time;
    model count/n = trt time trt*time / ddfm=kr;
    random intercept / subject=block;
    random _residual_ / subject=block*trt type=cs;
    lsmeans trt*time / ilink;
run;
```

- The conditional model (GLMM independent error form) does not use a working correlation matrix and does not have a *random* statement with the *_residual_* keyword.
- The marginal GEE-type model with a compound symmetry covariance structure uses a working correlation matrix specified by a *random* statement with the *_residual_* keyword in GLIMMIX. It is not a true GEE model because it contains a random block effect, but it is like a GEE in that it has a working correlation matrix.

Unlike the normal distribution case, the independent error and compound symmetry logit models produce very different results. The issues involved are similar to those discussed in Example 5.2.

The *Dimensions* sections for the two models are shown in Fig. 5–76 and 5–77. In the independent errors form, the G-side covariance parameters are the block and whole plot error variances. In contrast, in the compound symmetry form, the block variance is the only G-side covariance parameter, and the R-side parameters are the whole plot error and the within subjects correlation coefficient. Since the whole plot errors are no longer part of the G-side covariance structure, the number of columns in the random effects design matrix **Z** is reduced from five to one.

The fit statistics (Fig. 5–78 and 5–79) cannot and should not be used for model comparisons. The pseudo-likelihood approximation is model dependent. The GLMM and GEE-type models use different pseudo-likelihood approximations, and comparing them is very much a case of comparing apples and oranges.

As indicated in the discussion of the *Dimensions* sections, the two models estimate a different set of variance and covariance components. The block and between subjects (whole plot error) variance estimates are shown in Fig. 5–80 for the independent errors

FIG. 5-76. GLIMMIX output containing the *Dimensions* section for the independent errors form of the split plot model for Example 5.16.

FIG. 5-77. GLIMMIX output containing the *Dimensions* section for the compound symmetry form of the split plot model for Example 5.16.

Dimensions	
G-side Cov. Parameters	2
Columns in X	20
Columns in Z per Subject	5
Subjects (Blocks in V)	5
Max Obs per Subject	12

Dimensions	
G-side Cov. Parameters	1
R-side Cov. Parameters	2
Columns in X	20
Columns in Z per Subject	1
Subjects (Blocks in V)	5
Max Obs per Subject	12

FIG. 5-78. GLIMMIX output containing the fit statistics for the independent errors form of the split plot model for Example 5.16.

FIG. 5-79. GLIMMIX output containing the fit statistics for the compound symmetry form of the split plot model for Example 5.16.

Fit Statistics	
−2 Res Log Pseudo-Likelihood	1.09
Generalized Chi-Square	70.07
Gener. Chi-Square / DF	1.46

Fit Statistics	
−2 Res Log Pseudo-Likelihood	−3.55
Generalized Chi-Square	83.21
Gener. Chi-Square / DF	1.73

GLMM model. In categorical data terms, the block variance can be interpreted as the variance of the logarithm of the odds among blocks and the between subjects variance as the variance among log odds ratios among blocks. For the compound symmetry GEE-type model, the block variance estimate in Fig. 5–81 is interpreted similarly. The extra parameter in the compound symmetry GEE-type model, labeled *Residual*, is a scale parameter that can be interpreted as an over-dispersion parameter as discussed in Section 5.5. One could interpret the fact that $\hat{\phi}$ = 1.733 as evidence of over-dispersion. Finally, the *cs* covariance parameter is $\hat{\rho}$ = 0.654, the within subjects estimated correlation. This is a working correlation, not a true correlation, so beware of attaching too literal an interpretation to this estimate.

The tests for the fixed effects are different, most strikingly for the within subjects effects of time and treatment × time. In this example, the time effects are very large and the treatment × time interaction is so negligible that the overall conclusions would not change, but it is easy to see that this would not be true in general.

Although the interaction is not significant in either model, the least squares means are presented in Fig. 5–82 and 5–83 to illustrate the differences. The estimated probabilities in the *Mean* and *Standard Error Mean* columns do not agree. The estimated probabilities for the compound symmetry model are shifted toward 0.50 relative to those from the independent errors GLMM. This is consistent with discrepancies one would expect for conditional versus marginal models when the probabilities are all greater than 0.50, as previously discussed in Section 5.3. The

FIG. 5-80. GLIMMIX output containing the covariance parameter estimates and tests for the fixed effects for the independent errors form of the split plot model for Example 5.16.

Covariance Parameter Estimates

Cov Parm	Subject	Estimate	Standard Error
Intercept	block	0.02411	0.02021
trt	block	0.01247	0.006914

Type III Tests of Fixed Effects

Effect	Num DF	Den DF	F Value	Pr > F
trt	3	12.3	4.91	0.0183
time	2	48	60.52	<0.0001
trt*time	6	48	0.51	0.8008

FIG. 5-81. GLIMMIX output containing the covariance parameter estimates and tests for the fixed effects for the compound symmetry form of the split plot model for Example 5.16.

Covariance Parameter Estimates

Cov Parm	Subject	Estimate	Standard Error
Intercept	block	0.02377	0.02011
cs	block*trt	0.6540	0.5263
Residual		1.7334	0.4334

Type III Tests of Fixed Effects

Effect	Num DF	Den DF	F Value	Pr > F
trt	3	11.89	4.88	0.0194
time	2	32.13	343.35	<0.0001
trt*time	6	32.13	0.28	0.9419

standard errors differ partly because the GEE-type estimates are shifted toward 0.50 but mostly because of the presence of a scale parameter in the compound symmetry model and its absence in the independent errors GLMM model.

The results from the fits of the two models could be made more comparable by forcing the over-dispersion parameter ϕ to equal one in the compound symmetry model. This would not normally be part of the analysis but it is included here to illustrate the effect of the scale parameter on the differences between the results. Forcing $\phi = 1$ can be accomplished by adding the statement

parms (1) (1) (0)/hold = 2;

FIG. 5-82. GLIMMIX output containing the least squares means for the fungicide treatment × time interaction effects for the independent errors form of the split plot model for Example 5.16.

trt*time Least Squares Means

Fungicide treatment	Time (days)	Estimate	Standard Error	DF	t Value	Pr > |t|	Mean	Standard Error Mean
DCAD	12	1.5093	0.1034	13.24	14.60	<0.0001	0.8190	0.01533
DCAD	20	1.2597	0.1011	12.12	12.45	<0.0001	0.7790	0.01741
DCAD	42	1.1912	0.1006	11.87	11.84	<0.0001	0.7670	0.01798
None	12	1.2559	0.1011	12.1	12.42	<0.0001	0.7783	0.01744
None	20	0.9481	0.09905	11.15	9.57	<0.0001	0.7207	0.01994
None	42	0.8434	0.09851	10.91	8.56	<0.0001	0.6992	0.02072
RT	12	1.3492	0.1019	12.48	13.24	<0.0001	0.7940	0.01666
RT	20	1.0172	0.09946	11.33	10.23	<0.0001	0.7344	0.01940
RT	42	0.9841	0.09926	11.24	9.91	<0.0001	0.7279	0.01966
TSX	12	1.4811	0.1031	13.09	14.36	<0.0001	0.8147	0.01556
TSX	20	1.1431	0.1002	11.69	11.40	<0.0001	0.7582	0.01838
TSX	42	1.0034	0.09934	11.28	10.10	<0.0001	0.7317	0.01950

FIG. 5-83. GLIMMIX output containing the least squares means for the fungicide treatment × time interaction effects for the compound symmetry form of the split plot model for Example 5.16.

trt*time Least Squares Means

Fungicide treatment	Time (days)	Estimate	Standard Error	DF	t Value	Pr > |t|	Mean	Standard Error Mean
DCAD	12	1.5047	0.1130	17.53	13.31	<0.0001	0.8183	0.01681
DCAD	20	1.2556	0.1081	15.29	11.62	<0.0001	0.7783	0.01865
DCAD	42	1.1873	0.1069	14.77	11.11	<0.0001	0.7662	0.01915
None	12	1.2528	0.1080	15.27	11.60	<0.0001	0.7778	0.01867
None	20	0.9455	0.1034	13.25	9.15	<0.0001	0.7202	0.02083
None	42	0.8410	0.1022	12.73	8.23	<0.0001	0.6987	0.02151
RT	12	1.3450	0.1097	16.03	12.26	<0.0001	0.7933	0.01799
RT	20	1.0137	0.1043	13.63	9.72	<0.0001	0.7337	0.02037
RT	42	0.9806	0.1038	13.44	9.44	<0.0001	0.7272	0.02060
TSX	12	1.4810	0.1125	17.3	13.16	<0.0001	0.8147	0.01698
TSX	20	1.1431	0.1062	14.46	10.77	<0.0001	0.7582	0.01946
TSX	42	1.0035	0.1041	13.57	9.64	<0.0001	0.7317	0.02044

immediately after the second *random* statement in the GLIMMIX compound symmetry fit in Fig. 5–75. The *parms* statement provides GLIMMIX with starting values for the covariance parameters given in parentheses in the order in which they appear in the *Covariance Parameter Estimates* section of the output. The *hold* = 2 option requires the second covariance parameter in the list (the scale parameter ϕ) to be set to the specified starting value of one and not be estimated. Note that the addition of the *parms* statement changed the order of the estimates (Fig. 5–81 and 5–84).

The resulting covariance parameter estimates and tests for the fixed effects are shown in Fig. 5–84. Note the impact of the constraint on the compound symmetry correlation parameter and the fixed effects tests. The estimated correlation is 0.902 versus 0.654 when the scale parameter was unconstrained. The interpretation is still problematic. The discrepancy between the independent errors GLMM tests and the compound symmetry GEE-type tests is greatly reduced, but still exists. ∎

Example 5.16 yields two important insights into repeated measures modeling with non-normal data. First, the GLMM and GEE-type approaches to repeated measures do not produce equivalent results. Second, given that they are not equivalent, they could possibly produce contradictory results, leaving the researcher asking which analysis to use. We investigate this question by turning our attention to more complex covariance models.

As with the normal theory split plot in time there are two ways to model more complex covariance structures, one using the R-side approach, the other a purely GLMM or G-side approach. In the R-side approach, the working correlation is embedded in the variance function, much as it was with the compound symmetry structure in Example 5.16. For example, to use an unstructured covariance model in the GEE-type or R-side approach in the previous example, the working correlation matrix would be

FIG. 5–84. GLIMMIX output containing the covariance parameter estimates and tests for the fixed effects for the compound symmetry form of the split plot model for Example 5.16 when the overdispersion (scale) parameter is constrained to equal one.

Covariance Parameter Estimates

Cov Parm	Subject	Estimate	Standard Error
Intercept	block	0.02353	0.01995
Variance	block*trt	1.0000	.
CS	block*trt	0.9015	0.5081

Type III Tests of Fixed Effects

Effect	Num DF	Den DF	F Value	Pr > F
trt	3	11.83	4.87	0.0197
time	2	48	60.14	<0.0001
trt*time	6	48	0.51	0.8011

$$\text{diag}\left[\sqrt{\pi_{ijk}(1-\pi_{ijk})}\right]\begin{bmatrix} \phi_1^2 & \phi_{12} & \phi_{13} \\ \phi_{12} & \phi_2^2 & \phi_{23} \\ \phi_{13} & \phi_{23} & \phi_3^2 \end{bmatrix}\text{diag}\left[\sqrt{\pi_{ijk}(1-\pi_{ijk})}\right]$$

where the diagonal terms ϕ_k^2 act as over-dispersion scale parameters and the off-diagonal terms ϕ_{km} act as working covariance parameters. As with any working correlation matrix, these parameters do not have interpretations per se, but do yield consistent estimators that account for serial correlation. The matrix, including the variance function and the working covariance matrix, characterizes within subject variation and is specified through the *random* statement with the _residual_ keyword.

For the GLMM or G-side approach, a within subjects term v_{ijk} would be added to the linear predictor; i.e., the linear predictor would be given by

$$\eta_{ijk} = \log\left(\frac{\pi_{ijk}}{1-\pi_{ijk}}\right) = \beta_0 + B_i + F_j + T_k + FT_{jk} + v_{ijk}, i=1,\ldots,5; j=1,\ldots,4; k=1,2,3$$

where the distribution of the v_{ijk} would be given by

$$\mathbf{v}_{ij} = \begin{bmatrix} v_{ij1} \\ v_{ij2} \\ v_{ij3} \end{bmatrix} \sim \text{MVN}\left(\begin{bmatrix} 0 \\ 0 \\ 0 \end{bmatrix}, \begin{bmatrix} \sigma_1^2 & \sigma_{12} & \sigma_{13} \\ \sigma_{12} & \sigma_2^2 & \sigma_{23} \\ \sigma_{13} & \sigma_{23} & \sigma_3^2 \end{bmatrix} \right)$$

The difference in these two approaches lies with the effects in the linear predictor. In the R-side approach, η depends only on the fixed effects and any G-side covariance parameters, for example, a random block effect. In the GLMM or G-side approach, the linear predictor also depends on the within subjects term. It is assumed that there is a process that is being driven by the fixed effects and serial correlation effects within subjects. This is a critical distinction; the G-side approach embeds serial correlation in the linear predictor. In that sense, it behaves like a normally distributed process.

In the R-side approach, a quasi-likelihood is formed by embedding the working correlation in the variance function. The result is not a true likelihood. It does not describe a probability process that could actually happen, but it does produce consistent estimates of marginal treatment means and associated statistics.

The GLMM approach is based on a true likelihood. It describes a probability process that could plausibly happen and one that is easily simulated. It produces conditional estimates of the treatment means and associated statistics. In addition, the GLMM approach can be used to construct defensible best linear unbiased predictors when needed.

EXAMPLE 5.17

This example is a continuation of Examples 5.15 and 5.16 in which an unstructured covariance matrix is assumed and models are fit using the GEE-type (R-side) and GLMM approaches.

The GLIMMIX statements to fit these models are shown in Fig. 5–85. There are two differences between these coded models. First, the *random* statement that defines the unstructured model for the R-side approach uses the keyword _residual_ that does not appear in the GLMM or G-side approach. Second, the R-side approach can be implemented only with the default pseudo-likelihood, whereas the GLMM approach can be implemented either with pseudo-likelihood or with the integral approximation methods, Laplace and quadrature. The importance of this latter point will become apparent later.

The fit statistics for the two approaches are shown in Fig. 5–86 and 5–87. As with the GEE-type and GLMM comparison in Example 5.16, the fit statistics provide no useful information. Each model's pseudo-likelihood is based on a different linear predictor with different random effects and different distributional assumptions. They are not comparable and should not be compared. The generalized chi-square produces a value equal to or very close to the degrees of freedom, so the generalized chi-square/df will always be one aside from possible negligible rounding error. This is an artifact of the pseudo-variance of the unstructured model. Hence, the generalized chi-square/df statistics serves no diagnostic purpose.

Figures 5–88 and 5–89 present the estimated covariance parameters for fitted models and tests for the fixed effects. The differences are striking. For the GEE-type fit, the working covariance parameters are actually scale parameters. The

FIG. 5–85. GLIMMIX statements to fit an unstructured covariance model using GEE-type and GLMM for Example 5.17.

```
title2 'GEE-TYPE (R-SIDE) WITH UNSTRUCTURED COVARIANCE MATRIX';
proc glimmix data=fung;
  class block trt time;
  model count/n = trt time trt*time / ddfm=kr;
  random intercept / subject=block;
  random _residual_ / subject=block*trt type=un;
run;


title2 'GLMM (G-SIDE) WITH UNSTRUCTURED COVARIANCE MATRIX';
proc glimmix data=fung;
  class block trt time;
  model count/n = trt time trt*time / ddfm=kr;
  random intercept / subject=block;
  random time / subject=block*trt type=un;
run;
```

FIG. 5-86. GLIMMIX output containing the fit statistics for an unstructured covariance model using a GEE-type approach for Example 5.17.

FIG. 5-87. GLIMMIX output containing the fit statistics for an unstructured covariance model using GLMM for Example 5.17.

Fit Statistics	
−2 Res Log Pseudo-Likelihood	−23.16
Generalized Chi-Square	48.00
Gener. Chi-Square / DF	1.00

Fit Statistics	
−2 Res Log Pseudo-Likelihood	−24.34
Generalized Chi-Square	48.05
Gener. Chi-Square / DF	1.00

FIG. 5-88. GLIMMIX output containing the covariance parameter estimates and tests for fixed effects for an unstructured covariance model using a GEE-type approach for Example 5.17.

Covariance Parameter Estimates			
Cov Parm	Subject	Estimate	Standard Error
Intercept	block	0.03321	0.02577
UN(1,1)	block*trt	0.9711	0.3700
UN(2,1)	block*trt	1.0794	0.4678
UN(2,2)	block*trt	2.0821	0.7569
UN(3,1)	block*trt	-0.09807	0.4959
UN(3,2)	block*trt	0.5819	0.7275
UN(3,3)	block*trt	3.6898	1.3209

Type III Tests of Fixed Effects				
Effect	Num DF	Den DF	F Value	Pr > F
Trt	3	14.02	5.61	0.0097
Time	2	15	79.22	<0.0001
trt*time	6	18.28	0.39	0.8728

diagonal terms are the over-dispersion scale parameters for each time of measurement; e.g., $un(1, 1) = 0.9711$ is the estimate of ϕ_1^2. Assuming that the unstructured working correlation is the best GEE-type model for these data, $un(2, 2)$ and $un(3, 3)$ could be interpreted as evidence of over-dispersion for the measurements at Days 20 and 42. The off-diagonal terms, for example $un(2, 1)$, are estimates of the off-diagonal scale parameters associated with serial covariance. As before, these estimates lack intrinsic interpretation.

There are slight discrepancies between the fixed effects tests for the two fitted models. These result from the fact that the GEE-type approach tests marginal means, whereas the GLMM tests population averaged means and that the covariance assumptions of the two models are quite different. Despite this, the difference in the F and p-values are small. One would draw essentially identical

FIG. 5-89. GLIMMIX output containing the covariance parameter estimates and tests for fixed effects for an unstructured covariance model using GLMM for Example 5.17.

Covariance Parameter Estimates

Cov Parm	Subject	Estimate	Standard Error
Intercept	block	0.03123	0.02412
UN(1,1)	block*trt	1.55E-24	.
UN(2,1)	block*trt	0.01639	0.003507
UN(2,2)	block*trt	0.01505	0.008150
UN(3,1)	block*trt	-0.00188	0.007244
UN(3,2)	block*trt	0.007614	0.009640
UN(3,3)	block*trt	0.03413	0.01670

Type III Tests of Fixed Effects

Effect	Num DF	Den DF	F Value	Pr > F
Trt	3	17.42	5.52	0.0076
Time	2	16.34	80.60	<0.0001
trt*time	6	20.02	0.50	0.8014

conclusions from either analysis. This is not necessarily true in general. Success probabilities closer to zero or one will accentuate the difference between marginal and conditional means. In addition, in this data set there are $n = 400$ Bernoulli trials per block–treatment–time combination, an unusually large number. A smaller number of trials would tend to amplify discrepancies between the GEE-type and GLMM results. ■

COMPARING COVARIANCE STRUCTURES

Examples 5.16 and 5.17 demonstrate that the fixed effects tests can be substantially affected by the covariance model. Similarly the covariance model affects estimates of the treatment and time effects (marginal and conditional) and their standard errors. The standard errors tend to be affected to a greater extent than the estimates. Clearly the choice of covariance model matters for non-normally distributed data just as it does for normally distributed data.

Unfortunately covariance model selection is not as straightforward for non-normal repeated measures data as it is under the normality assumption. As was seen in the examples, the fit statistics associated with pseudo-likelihood estimation are not comparable among models and, hence, cannot be used to choose between competing covariance structures. For the GEE-type approach, because they are based on quasi-likelihood theory, there is no formal statistical procedure for covariance model selection. For the GLMM approach, the situation is better.

As indicated previously, the GLMM defines an actual likelihood, a true probability process. While GLIMMIX's default computing algorithm for GLMMs is

180   CHAPTER 5

pseudo-likelihood, it also allows two alternative methods that work directly with the actual likelihood, namely, the Laplace and Gauss–Hermite quadrature methods. These methods are computationally more involved than pseudo-likelihood but because the actual likelihood is evaluated, meaningful likelihood ratio tests can be constructed and comparable information criteria can be computed. Hence, for GLMM G-side covariance structures, covariance model selection can proceed much as it does for normally distributed data as long as either the Laplace or quadrature methods are used. Of these two, the Laplace method is preferred because quadrature is usually computationally prohibitive for typical repeated measures GLMMs.

EXAMPLE 5.18

This example is a continuation of the previous examples in this section in which a series of covariance structures using the GLMM or G-side approach are compared to determine an appropriate model for the data.

Figure 5–90 shows the GLIMMIX statements to fit G-side repeated measures models. The statements show the compound symmetry model, but the same modifications work for all G-side models. The difference between this and the corresponding program in Fig. 5–85 is the *method* option on the PROC GLIMMIX statement and the *type* option on the second *random* statement.

The fit statistics for compound symmetry are shown in Fig. 5–91. These statistics are based on the full likelihood, not the residual likelihood as in the REML estimation of covariance models for normally distributed data. The full likelihood includes fixed effects (treatment and time) and random effects (block and within-subjects effects). When selecting a covariance structure, models for which the only change is in the covariance structure model can be legitimately compared using these fit statistics.

The model was fit to several of the covariance structures that were described in Section 4.6 and used in Example 4.5 for a normally distributed response. The Laplace method was used in all of the fits and different covariance structures were obtained by changing the *type* option. The AICC values for the current data are shown in Table 5–2. Because a smaller value of AICC indicates a better fit, the first order autoregressive model appears to be the model of choice.

The GLIMMIX statements to fit the *ar(1)* model are shown in Fig. 5–92. The covariance parameter estimates and test of a non-zero autocorrelation coefficient

FIG. 5–90. GLIMMIX statements to fit a compound symmetry covariance model using the G-side approach for Example 5.18.

```
title2 'GLMM (G-SIDE) WITH COMPOUND SYMMETRY COVARIANCE STRUCTURE';
proc glimmix data=fung  method=laplace;
  class block trt time;
  model count/n = trt time trt*time;
  random intercept / subject=block;
  random time / subject=block*trt type=cs;
run;
```

FIG. 5-91. GLIMMIX output containing the fit statistics for the G-side approach with a compound symmetry covariance structure for Example 5.18.

Fit Statistics	
–2 Log Likelihood	468.48
AIC (smaller is better)	498.48
AICC (smaller is better)	509.39
BIC (smaller is better)	492.62
CAIC (smaller is better)	507.62
HQIC (smaller is better)	482.75

TABLE 5-2. Small sample corrected Akaike information criteria (AICC) for selected covariance structures for the within subjects model in Example 5.18. Smaller AICC values indicate more parsimonious models.

Covariance structure	GLIMMIX type option	AICC
Unstructured	un	511.62
Compound symmetry	cs	509.39
Heterogeneous compound symmetry	csh	512.03
Independence with between subjects effect	–	508.67
First order autoregressive	ar(1)	507.79
Heterogeneous AR(1)	arh(1)	513.39
Heterogeneous Toeplitz	toeph	516.17
First order ante-dependence	ante(1)	507.87

are shown in Fig. 5–93. The tests for the fixed effects are shown in Fig. 5–94. The treatment × time interaction is not significant, but both the treatment and time main effects are significant at the 0.05 level. The remainder of the analysis would involve the least squares means for both main effects. ∎

5.9 INFERENCE ISSUES FOR REPEATED MEASURES GENERALIZED LINEAR MIXED MODELS

In repeated measures with normally distributed responses, two inference issues, standard error bias and denominator degrees of freedom, motivated the use of the Kenward–Roger adjustment as a recommended standard operating procedure (Section 4.2). The theoretical basis for this adjustment is REML estimation for the normal distribution. In its literal form, this theoretical basis disappears for non-normal generalized linear mixed models. Pseudo-likelihood estimation mimics REML using a pseudo-variable based on the link function. The pseudo-likelihood

FIG. 5-92. GLIMMIX statements to fit a first order autoregressive covariance model using GLMM for Example 5.18.

```
title2 'GLMM (G-SIDE) WITH AR(1) COVARIANCE STRUCTURE';
proc glimmix data=fung method=laplace;
   class block trt time;
   model count/n = trt time trt*time;
   random intercept / subject=block;
   random time / subject=block*trt type=ar(1);
   covtest 'rho = 0' . . 0 ;
run;
```

FIG. 5-93. GLIMMIX output containing the covariance parameter estimates and test for the auto-correlation coefficient equal to zero for the first order autoregressive model in Example 5.18.

Covariance Parameter Estimates

Cov Parm	Subject	Estimate	Standard Error
Intercept	block	0.01955	0.01470
Variance	block*trt	0.01334	0.005773
AR(1)	block*trt	0.6397	0.2256

Tests of Covariance Parameters Based on the Likelihood

Label	DF	−2 Log Like	ChiSq	Pr > ChiSq	Note
rho = 0	1	471.78	4.90	0.0269	DF

DF: P-value based on a chi-square with DF degrees of freedom.

FIG. 5-94. GLIMMIX output containing the tests for the fixed effects for the first order autoregressive model in Example 5.18.

Type III Tests of Fixed Effects

Effect	Num DF	Den DF	F Value	Pr > F
trt	3	44	5.97	0.0017
time	2	44	38.33	<0.0001
trt*time	6	44	0.33	0.9186

is essentially what the function would be if the pseudo-variable was in fact a real, normally distributed variable. The Kenward–Roger computing formulas can be implemented accordingly. In this sense, they are pseudo-Kenward–Roger adjustments. This is not necessarily bad. Simulation studies suggest that when the approximation used to create the pseudo-variable is valid, as it is for a binomial with reasonably large n as in the examples in Section 5.8, the pseudo-likelihood procedure in conjunction with the Kenward–Roger adjustment is quite accurate.

When the Laplace and quadrature methods are used the situation is different. The likelihood is evaluated directly by an integral approximation. There is no pseudo-variable and hence, no residual pseudo-likelihood and no basis for computing a Kenward–Roger adjustment. When these methods are used in GLIMMIX, the *kr* option is no longer available. However, the standard error bias issue remains. Zeger and Liang (1986) suggested using a "sandwich estimator," also referred to as an empirical or robust estimator in the generalized linear mixed model literature.

In GLIMMIX the *empirical* option on the *proc* statement causes the sandwich estimator to be used for all test statistics and standard errors. The *empirical* option can be used with the default pseudo-likelihood or quadrature methods.

EXAMPLE 5.19

This example is a continuation of Example 5.18 in which the first order autoregressive model was chosen as the most appropriate covariance structure model based on AICC. In this example, the model is fit using the G-side approach and different standard error adjustments.

Figures 5–95 and 5–96 show the results of the fixed effects tests from the fitted models using the Kenward–Roger adjustment with pseudo-likelihood estimation and the sandwich estimator with Laplace estimation, respectively. The outputs reveal a problem with the sandwich estimator. It appears to be severely biased for small sample sizes. Most agronomic experiments have only a few replications, typically 3 to 6, depending on practical restrictions or what is required for

FIG. 5–95. GLIMMIX output containing the tests for the fixed effects using pseudo-likelihood with the Kenward–Roger adjustment for the ar(1) model in Example 5.19.

Type III Tests of Fixed Effects

Effect	Num DF	Den DF	F Value	Pr > F
trt	3	12.84	4.92	0.0172
time	2	41.11	30.00	<0.0001
trt*time	6	41.17	0.26	0.9525

FIG. 5–96. GLIMMIX output containing the tests for the fixed effects using the Laplace method with the sandwich estimator adjustment for the ar(1) model in Example 5.19.

Type III Tests of Fixed Effects

Effect	Num DF	Den DF	F Value	Pr > F
trt	3	44	10.16	<0.0001
time	2	44	319.55	<0.0001
trt*time	5	44	718685	<0.0001

FIG. 5-97. GLIMMIX output containing the tests for the fixed effects using the Laplace method with the Morel bias corrected sandwich estimator adjustment for the ar(1) model in for Example 5.19.

Type III Tests of Fixed Effects				
Effect	Num DF	Den DF	F Value	Pr > F
trt	3	44	2.25	0.0961
time	2	44	24.42	<0.0001
trt*time	6	44	0.10	0.9959

adequate power. The sandwich estimator is best suited to large-scale studies such as clinical trials that typically have hundreds or even thousands of subjects.

There are a number of bias-adjusted sandwich estimators, most notably a procedure by Morel et al. (2003) that can be used in GLIMMIX. The Morel correction is invoked by including the *empirical* = *mbn* option on the PROC GLIMMIX statement. The bias-corrected fixed effects tests are presented in Fig. 5–97. These results are conservative relative to the *kr*-adjusted statistics obtained from pseudo-likelihood estimation. While not shown here, the *mbn*-corrected empirical standard errors are extremely conservative relative to the *kr*-computed standard errors. ∎

This issue is in need of more study. The *mbn* correction has shown promise in the pharmaceutical applications for which it was developed. Whether its promise holds for applications that agronomic researchers face is an unanswered question at this time.

5.10 MULTINOMIAL DATA

A multinomial model is the generalization of the binomial model to more than two categories. The multinomial categories can be nominal (unordered) or ordinal (ordered). For example, in a study of red rice genetics, hull color (straw, brown, black) would form a set of nominal categories. Visual ratings of the strength of a set of cultivars' resistance to a crop disease or the amount of weed control by a herbicide at various rates could be represented by a set of ordinal categories. In each case, the response variables would be a set of counts of the number of items in each category.

Consider a simple example of visual disease ratings with four ordinal categories—none, slight, moderate, and severe—from an experiment containing several treatments. Data of this type have often been analyzed by assigning a numeric code to the ordered categories and treating the resulting codes as if they had a normal distribution. For example, the ratings might be coded as 0 (none), 1 (slight), 2 (moderate), and 3 (severe). The end result of such analyses is a treatment mean or treatment difference. Suppose that the results of such an analysis yielded the mean of treatment A as 1.5 (presumably meaning "slight to moderate") and the mean of treatment B as 1.75 (perhaps meaning "more moderate than slight"). What interpretation can be given to the difference in treatment means, 0.25? Would the difference be interpreted similarly if the treatment means were 3.0 and 2.75 or if they were 0.5 and 0.25, respectively? Obviously, interpretation of such analyses is problematic. Treatment A could have a mean of 1.5 because half of the observations had slight symptoms and half had moderate symptoms. On the other hand, treatment A could also have a mean of 1.5 because half of the observations were rated none and half were rated severe. The same numerical value of the mean would appear to have very different meanings. How one interprets a difference of 0.25 is an open question with no satisfactory answer.

A better approach is to model such data using multinomial generalized linear mixed models. These are essentially extensions of logit and probit models for binomial data, except that they apply to three or more response categories instead of the binomial's two categories.

In this section, we give a brief conceptual description of the multinomial generalized linear model followed by an example. The focus will be on ordinal data since our experience is that the majority of agricultural categorical data is ordinal. The concepts presented here can be extended to non-ordinal data.

CONCEPTUAL DESCRIPTION

Recall that for binomial data, the fundamental idea is that the predictors, be they treatment effects, regression effects, or block effects, directly affect the link function η, but we can only observe the consequence of a particular value of η and not the link function itself. More specifically, the link function determines the probability of observing either a success or failure. For example, in a logit model, the model predictors determine η, which in turn determines the probability $\pi = 1/(1 + e^{-\eta})$. Changing the model predictors changes η and, hence, the probability $\pi = 1/(1 + e^{-\eta})$ of observing a success.

In a probit model, one can think of η as an unobservable normally distributed random variable. When η is above or below some threshold, we observe either a 0 ("failure") or 1 ("success"), resulting in a binomially distributed observed response.

Both the logit and probit inverse link functions describe a cumulative probability that depends on η. Figure 5–98 illustrates this idea. The figure can be viewed as the inverse link of either the logit or the normal cumulative probability function since the shapes of these two functions are very similar. Since η is on the horizontal axis and π is on the vertical axis, the function that relates η to π is the inverse link.

Ordinal multinomial models extend the logit and probit concept. Instead of dividing the η scale (horizontal axis) into two segments, one representing failure and the other representing success, the η scale is divided into three or more segments, one for each response category. Figure 5–99 illustrates the idea. The category boundaries on the link scale are denoted by η_1 and η_2. If the observed values of the predictors yield a value of the link η less than η_1, then category 1 is observed. If the observed value of η is greater than η_2, then category 3 is observed. Category 2 is observed if η lies between η_1 and η_2.

As in the binomial model, the category boundaries depend on the treatment effects. For example, for the three category ordinal multinomial model, suppose that the experiment contains only two treatments; e.g., treated and untreated. Let T_i, for $i = 1, 2$, represent the treatment effects. For the ith treatment, the boundary

FIG. 5-98. Binomial probabilities as a function of the link. The vertical dashed line and its projection to the cumulative probability axis indicate a threshold defining two categorical outcomes.

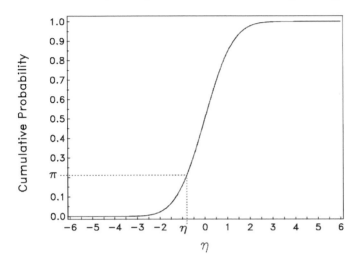

FIG. 5-99. Cumulative probability for a three category ordinal multinomial model as a function of the link. The vertical dashed lines and their projections to the cumulative probability axis indicate thresholds defining three categorical outcomes.

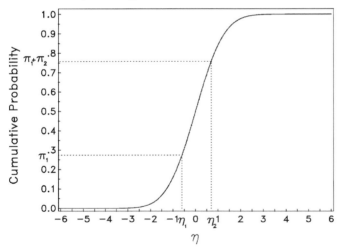

between c and $c + 1$ is given by $\eta_c = \beta_{0c} + T_i$, where β_{0c} is the intercept for the cth boundary between categories. There will be one value of the link function for each boundary. For both the cumulative logit and probit, the two linear predictors will be $\eta_{1i} = \beta_{01} + T_i$ and $\eta_{2i} = \beta_{02} + T_i$. The first linear predictor η_{1i} describes how the lower dividing boundary varies with the treatment and the second linear predictor η_{2i} describes how the upper dividing boundary varies. Note that the predictors have different intercepts and that $\beta_{01} < \beta_{02}$. The model assumes that the treatment affects both linear predictors in the same way; i.e., only through the treatment effect T_i. Changing treatments affects both boundaries on the link scale η by the difference $T_1 - T_2$. When the treatment varies, both linear predictors vary in such a way that the distance between them on the horizontal axis remains constant.

More generally, the cumulative logit and probit models can be used for any mixed model. For both models, the general form of the linear predictors is

$\eta_{1i} = \beta_{01} +$ sum of fixed effects $+$ sum of random effects,

$\eta_{2i} = \beta_{02} +$ sum of fixed effects $+$ sum of random effects.

In general, if the response variable has c categories, there will be $c - 1$ linear predictors. For example, if there are five response categories, there are four linear predictors. Each linear predictor will have a unique intercept, but all other fixed and random effect terms in the model are shared.

To relate the linear predictors to the multinomial distribution, let π_j be the probability of an observation being in the jth response category. In Fig. 5–99, there are three categories and, hence, three probabilities. Since the purpose of the model is to determine the effect of treatment on the response probability, let π_{ji} denote the probability of an observation receiving the ith treatment falling into the jth response category.

The relationship between the link functions and the multinomial probabilities can be visualized using Fig. 5–99. The linear predictors $\eta_{1i}, \eta_{2i}, \ldots$ defined above are the category boundaries at the ith level of the treatment (or predictor) variable; η_{1i} is the lowest boundary, η_{2i} is the next lowest, etc. The corresponding values on the cumulative probability scale are used to obtain the estimated multinomial probabilities.

CUMULATIVE PROBIT (THRESHOLD) MODEL

For the cumulative probit or threshold model, the inverse links and cumulative probabilities can be visualized in Fig. 5–99 as follows:

$\pi_{1i} = \Phi(\eta_{1i})$

where $\Phi(\eta_{1i})$ is the area under the normal curve (probability) up to the lower boundary defined by η_{1i},

$\pi_{2i} = \Phi(\eta_{2i}) - \Phi(\eta_{1i})$

where $\Phi(\eta_{2i})$ is the area under the normal curve up to the upper boundary defined by η_{2i}. Subtracting $\pi_{1i} = \Phi(\eta_{1i})$ yields the area under the normal curve for category 2 only.

$\pi_{3i} = 1 - \Phi(\eta_{2i})$

where subtraction yields the area under the normal curve above the upper category boundary.

PROPORTIONAL ODDS MODEL

The proportional odds model uses cumulative logits. The link functions are the logits of the cumulative probabilities up to and including the cth category. The first linear predictor models the logit of π_1, where category $j = 1$ is the lowest ordinal category. The second linear predictor models the logit of the cumulative probability of the two lowest categories, $\pi_1 + \pi_2$, etc.

For a three category proportional odds linear regression model, the linear predictors would be given by

$$\eta_{1i} = \log\left(\frac{\pi_{1i}}{1 - \pi_{1i}}\right) = \beta_{01} + \beta_1 T_i$$

$$\eta_{2i} = \log\left[\frac{\pi_{1i} + \pi_{2i}}{1 - (\pi_{1i} + \pi_{2i})}\right] = \beta_{02} + \beta_1 T_i$$

Once the model has been fitted, the inverse links can be used to obtain estimates of the response probabilities. The inverse links are given by

$$\pi_{1i} = \frac{1}{1 + e^{-\eta_{1i}}}$$

$$\pi_{2i} = \frac{1}{1 + e^{-\eta_{2i}}} - \frac{1}{1 + e^{-\eta_{1i}}}$$

$$\pi_{3i} = 1 - \frac{1}{1 + e^{-\eta_{2i}}}$$

For both the cumulative probit and proportional odds multinomial models, when the predictor changes (e.g., if the treatment changes), the boundary points on the η axis move, thereby changing the cumulative probabilities. Using the one factor, two treatment (treated, untreated) example discussed previously, suppose there is a treatment effect that has a positive effect on the linear predictor η. In addition, suppose that Fig. 5–99 represents the untreated group and Fig. 5–100 represents the treated group. The boundaries in Fig. 5–100 are shifted to the right by $T_2 - T_1$ since the effect on η was positive. The cumulative probabilities are affected accordingly. It is important to understand that when modeling multinomial data the change in the predictor variable affects all linear predictors (i.e., the boundary points) simultaneously and in the same way on the η axis. In turn, this affects the probabilities of all categories simultaneously according to the shape of the inverse link function, which in general, is not linear.

FIG. 5-100. Cumulative probabilities for a three category ordinal multinomial model for the treated group in a two treatment experiment. The vertical dashed lines and their projections to the cumulative probability axis indicate thresholds defining three categorical outcomes.

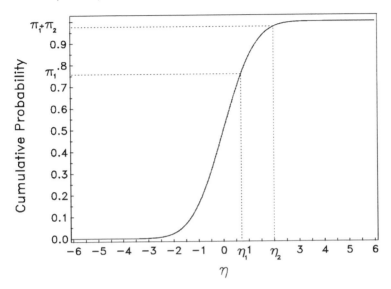

The assumption that all linear predictors are equally affected by all model effects and, hence, that the distances between category boundaries remain constant as the effects change may not be realistic in some applications. While there are more advanced models that relax this assumption, they can be quite complex and are not considered here.

EXAMPLE 5.20

Stink bugs are an insect pest that feeds on the pods of soybean plants. Damage is caused when they insert their piercing and sucking mouth parts into the seedpods to feed on plant juices. Their digestive juices lead to plant tissue degeneration. Stink bug feeding during pod development results in shriveled seeds, reduced seed size, seed discoloration, and lower seed quality.

Data for this example are part of an unpublished study conducted by C. Capps (used with permission) to compare the potential damage caused by two stink bug species, Southern green and Red-banded. Thirty soybean plants in a field were randomly assigned to one of the two species or used as an untreated control, with 10 plants per treatment. Individual plants were caged and stink bug nymphs were placed in each cage. The cages were removed after 10 days, and the plants were treated with insecticide as needed to prevent further damage. At harvest, all pods from the plants were removed, and seeds were examined. Individual seeds were visually rated for shriveling (*shrivel*) using the rating scale in Table 5–3. The number of seeds per plant ranged from 21 to 48. Only data from the seed closest to the end of the pod where it was attached to the plant are included here.

The species treatments (*species*) were a fixed effect, and the plants (*plant*) were a random effect. Since the visual ratings are ordinal, a proportional odds multinomial model was fit to the data.
Let

Y_{1ij} = number of seeds rated 1 from the *j*th plant assigned to the *i*th species,

...

Y_{5ij} = number of seeds rated 5 from the *j*th plant assigned to the *i*th species,

TABLE 5-3. Rating categories for visual evaluation of percent shriveling for Example 5.20.

Rating category	Visual evaluation of percentage shriveling
	— % —
1	0–5
2	5–25
3	25–50
4	50–75
5	75–100

where $i = 1, 2, 3$ and $j = 1, 2, ..., 10$,

Let $\pi_{1ij}, \pi_{2ij}, \pi_{3ij}, \pi_{4ij}$ and π_{5ij} be the probabilities that a seed will be rated 1, 2, 3, 4, or 5, respectively. Assume that $[Y_{1ij}, Y_{2ij}, Y_{3ij}, Y_{4ij}, Y_{5ij}]'$ has a multinomial distribution for each *i* and *j*. Then the four linear predictors are given by

$$\eta_{1ij} = \log\left(\frac{\pi_{1ij}}{1 - \pi_{1ij}}\right) = \beta_{01} + S_i + P_{j(i)}$$

$$\eta_{2ij} = \log\left[\frac{\pi_{1ij} + \pi_{2ij}}{1 - (\pi_{1ij} + \pi_{2ij})}\right] = \beta_{02} + S_i + P_{j(i)}$$

$$\eta_{3ij} = \log\left[\frac{\pi_{1ij} + \pi_{2ij} + \pi_{3ij}}{1 - (\pi_{1ij} + \pi_{2ij} + \pi_{3ij})}\right] = \beta_{03} + S_i + P_{j(i)}$$

$$\eta_{4ij} = \log\left[\frac{\pi_{1ij} + \pi_{2ij} + \pi_{3ij} + \pi_{4ij}}{1 - (\pi_{1ij} + \pi_{2ij} + \pi_{3ij} + \pi_{4ij})}\right] = \beta_{04} + S_i + P_{j(i)}$$

where β_{0c} is the intercept for the *c*th predictor, S_i is the *i*th species effect, and $P_{j(i)}$ is the effect of the *j*th plant assigned to the *i*th species. Note that the species and plant effects are the same for all four linear predictors; only the intercepts differ.

The GLIMMIX statements to fit the model are shown in Fig. 5–101. The data set *stinkbug* contains one line per seed. The *model* statement specifies the multinomial distribution and the cumulative probabilities in the logit link function. Since the *empirical* option specifies the sandwich estimator adjustment for the covariance parameter, the *random* statement uses the *subject* option format.

The basic model and fitting information are shown in Fig. 5–102. The 1028 observations represent the number of seeds examined. The *Response Profile* shows the shriveling rating codes, the order used by GLIMMIX, the number of seeds receiving each rating, and the order of the categories used by GLIMMIX. The G-side covariance parameter is the variance of the random plant effect. The columns in the fixed effects design matrix **X** correspond to the four intercept terms and the three species treatment effects.

The estimated plant variance and the test for the fixed species effect are presented in Fig. 5–103. There are significant species treatment differences.

The *estimate* statements in Fig. 5–101 determine the estimates of the linear predictors, and the *ilink* option converts them to the cumulative probability scale. The results are shown in Fig. 5–104. The estimates of the linear predictors, η_{ci}, are found under the column labeled *Estimate* and the corresponding estimated cumulative probabilities under *Mean*. For example, in second line of the table labeled *shrivel = 2, redband*, the estimated linear predictor representing the boundary between shrivel categories 2 and 3 for Red-banded stink bugs is 0.1979. The estimate of the corresponding cumulative probability for shrivel categories 1 and 2, $\pi_{11} + \pi_{21}$, is 0.5493. Since the probability for category 1, π_{11}, is given in the first line of the table as 0.3966, the estimated probability for category 2 is 0.5493 – 0.3966 = 0.1527; i.e., the probability a seed being classified as shriveling category 2 (5–25% shriveling) when attacked by the Red-banded species is 0.1527.

Using the estimates in Fig. 5–104, the probability of each response category for each species was calculated and is given in Table 5–4. The estimated linear

FIG. 5–101. GLIMMIX statements to fit the proportional odds multinomial model for Example 5.20.

```
proc glimmix data=stinkbug method=laplace empirical=mbn;
  class species plant;
  model shrivel = species / dist=multinomial link=cumlogit;
  random intercept / subject=plant(species);
  estimate 'shrivel=1, redband' intercept 1 0 0 0 species 1 0 0 / ilink;
  estimate 'shrivel=2, redband' intercept 0 1 0 0 species 1 0 0 / ilink;
  estimate 'shrivel=3, redband' intercept 0 0 1 0 species 1 0 0 / ilink;
  estimate 'shrivel=4, redband' intercept 0 0 0 1 species 1 0 0 / ilink;

  estimate 'shrivel=1, southgreen' intercept 1 0 0 0 species 0 1 0 / ilink;
  estimate 'shrivel=2, southgreen' intercept 0 1 0 0 species 0 1 0 / ilink;
  estimate 'shrivel=3, southgreen' intercept 0 0 1 0 species 0 1 0 / ilink;
  estimate 'shrivel=4, southgreen' intercept 0 0 0 1 species 0 1 0 / ilink;

  estimate 'shrivel=1, untreated' intercept 1 0 0 0 species 0 0 1 / ilink;
  estimate 'shrivel=2, untreated' intercept 0 1 0 0 species 0 0 1 / ilink;
  estimate 'shrivel=3, untreated' intercept 0 0 1 0 species 0 0 1 / ilink;
  estimate 'shrivel=4, untreated' intercept 0 0 0 1 species 0 0 1 / ilink;

  contrast 'redband vs southgreen' species 1 -1 0;
  contrast 'untreated vs species avg' species 1 1 -2;
run;
```

FIG. 5-102. GLIMMIX output containing basic model and fitting information for Example 5.20.

Model Information

Data Set	WORK.STINKBUG
Response Variable	Shrivel
Response Distribution	Multinomial (ordered)
Link Function	Cumulative Logit
Variance Function	Default
Variance Matrix Blocked By	Plant(Species)
Estimation Technique	Maximum Likelihood
Likelihood Approximation	Laplace
Degrees of Freedom Method	Containment
Fixed Effects SE Adjustment	Sandwich - MBN(df,r=1,d=2)

Class Level Information

Class	Levels	Values
Species	3	Redbanded SoGreen Untreated
Plant	10	1 2 3 4 5 6 7 8 9 10

Number of Observations Read	1028
Number of Observations Used	1028

Response Profile

Ordered Value	Shrivel	Total Frequency
1	1	600
2	2	121
3	3	62
4	4	31
5	5	214

The GLIMMIX procedure is modeling the probabilities of levels of Shrivel having lower Ordered Values in the Response Profile table.

Dimensions

G-side Cov. Parameters	1
Columns in X	7
Columns in Z per Subject	1
Subjects (Blocks in V)	30
Max Obs per Subject	48

predictors and cumulative probabilities for each species treatment are plotted in Fig. 5–105. Based on the table and graphs, it appears that there may not be a difference between the damage caused by the two species, but they appear to be very

FIG. 5-103. GLIMMIX output containing the covariance parameter estimate and the test for the fixed effect for Example 5.20.

Covariance Parameter Estimates

Cov Parm	Subject	Estimate	Standard Error
Intercept	Plant(Species)	0.1345	0.1135

Type III Tests of Fixed Effects

Effect	Num DF	Den DF	F Value	Pr > F
Species	2	27	17.86	<0.0001

FIG. 5-104. GLIMMIX output containing the estimated linear predictors on the cumulative logit scale and the cumulative probabilities for Example 5.20.

Estimates

Label	Estimate	Standard Error	DF	t Value	Pr > |t|	Mean	Standard Error Mean
shrivel=1, redband	-0.4198	0.2658	27	-1.58	0.1260	0.3966	0.06362
shrivel=2, redband	0.1979	0.2376	27	0.83	0.4121	0.5493	0.05882
shrivel=3, redband	0.5481	0.2277	27	2.41	0.0232	0.6337	0.05284
shrivel=4, redband	0.7375	0.2256	27	3.27	0.0029	0.6765	0.04937
shrivel=1, southgreen	0.2293	0.1935	27	1.18	0.2465	0.5571	0.04775
shrivel=2, southgreen	0.8470	0.1818	27	4.66	<0.0001	0.6999	0.03819
shrivel=3, southgreen	1.1972	0.1902	27	6.29	<0.0001	0.7680	0.03389
shrivel=4, southgreen	1.3866	0.2004	27	6.92	<0.0001	0.8000	0.03206
shrivel=1, untreated	1.5655	0.2398	27	6.53	<0.0001	0.8271	0.03428
shrivel=2, untreated	2.1832	0.2397	27	9.11	<0.0001	0.8987	0.02182
shrivel=3, untreated	2.5334	0.2763	27	9.17	<0.0001	0.9265	0.01883
shrivel=4, untreated	2.7228	0.2958	27	9.21	<0.0001	0.9384	0.01711

different from the untreated controls, which represent the effect of the underlying field infestation.

The two hypotheses in the previous paragraph were tested using the *contrast* statements in Fig. 5–101. The results shown in Fig. 5–106 provide evidence that the untreated effect differs from the average of the two species treatments and that there is a statistically significant difference between the effects of the Red-banded and Southern green species. Note that these tests apply to the species effects on the cumulative logit scale. In addition, when the species effects are statistically significant, the interpretation is that they affect all of the category boundaries as a group.

TABLE 5-4. Estimated probabilities of shriveling damage by species treatment calculated from the estimates in Fig. 5–104 for Example 5.20.

		Estimated probability		
Shriveling rating	Percentage shriveling	Red-banded	Southern green	Untreated
	— % —			
1	0–5	0.3966	0.5571	0.8271
2	5–25	0.1527	0.1428	0.0716
3	25–50	0.0844	0.0681	0.0278
4	50–75	0.0428	0.0320	0.0119
5	75–100	0.3235	0.2000	0.0616

While this is a test on the logit scale, the impact on the probabilities of the various categories follows automatically, as depicted graphically in Fig. 5–105.

If an underlying cumulative probit (threshold) model had been assumed instead of the cumulative logit (proportional odds) model, similar interpretations would have been made. In both types of models, applying a separate conclusion to treatment effects on individual categories has no meaning; that is, treatments affect the collection of probabilities as a group, not individually.

Finally, the focus of the proportional odds model is on the probabilities of the various response categories for each treatment rather than the average rating obtained under the assumption that the seed counts were normally distributed. ∎

FIG. 5-105. Graphs of estimated cumulative probabilities as a function of the linear predictors by species for Example 5.20. Red-banded (top), Southern green (middle), Untreated (bottom). The vertical dashed lines and their projections to the cumulative probability axis indicate thresholds defining the categorical outcomes.

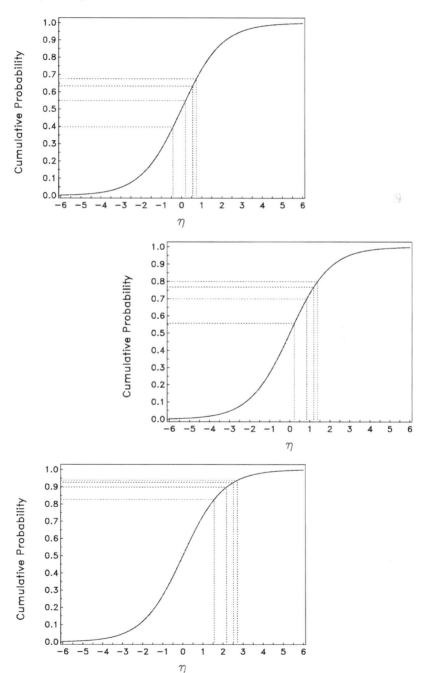

FIG. 5-106. GLIMMIX output containing the tests for the contrasts for Example 5.20.

Contrasts				
Label	Num DF	Den DF	F Value	Pr > F
redband vs southgreen	1	27	5.08	0.0326
untreated vs species avg	1	27	33.62	<0.0001

REFERENCES CITED

Ehlenfeldt, M.K., J.J. Polashock, A.W. Stretch, and M. Kramer. 2010. Mummy berry fruit rot and shoot blight incidence in blueberry: Prediction, ranking, and stability in a long-term study. HortScience 45:92–97.

Faraway, J.J. 2006. Extending the linear model with R: Generalized linear, mixed effects and nonparametric regression models. Chapman and Hall, CRC Press, Boca Raton, FL.

Hardin, J.W., and J.M. Hilbe. 2003. Generalized estimating equations. Chapman and Hall, CRC Press, Boca Raton, FL.

McCullagh, P., and J.A. Nelder. 1989. Generalized linear models. 2nd ed. Chapman and Hall, New York.

McLean, R.A., W.L. Sanders, and W.W. Stroup. 1991. A unified approach to mixed linear models. Am. Stat. 45:54–64. doi:10.2307/2685241

Molenberghs, G., and G. Verbeke. 2006. Models for discrete longitudinal data. Springer Verlag, New York.

Morel, J.G., M.C. Bokossa, and N.K. Neerchal. 2003. Small sample correction for the variance of GEE estimators. Biometric. J. 45:395–409. doi:10.1002/bimj.200390021

Stroup, W.W., and J. Stubbendieck. 1983. Multivariate statistical methods to determine changes in botanical composition. J. Range Manage. 36:208–212. doi:10.2307/3898164

Zeger, S.L., and K.-Y. Liang. 1986. Longitudinal data analysis for discrete and continuous outcomes. Biometrics 42:121–130. doi:10.2307/2531248

MORE COMPLEX EXAMPLES

6.1 INTRODUCTION

In previous chapters the numerical examples were used to illustrate specific aspects of generalized linear mixed models and their analysis. In this chapter, two generalized linear mixed model examples of more complex experimental situations are presented. Both examples involve modeling the correlation structure of random effects and/or residuals.

6.2 REPEATED MEASURES IN TIME AND SPACE

In designed studies, the response variable may be measured more than once on the same experimental unit. As discussed previously in Sections 4.5, 4.7, 5.8 and 5.9, these multiple measurements from the same experimental unit are referred to as repeated measures. Often such measurements are made over time. For example, plant height (the response variable) may be recorded weekly throughout the growing season. The observations from the same plot would be anticipated to be correlated, with the correlation being stronger among observations closer together in time than those further apart. Repeated measures may also be collected over space. As an example, soil cores may be taken from a plot, and measurements made at a series of depths. Within a soil core, measurements from two depths closer together would likely be more similar than those further apart. The first example in this chapter involves both types of repeated measures, one in space and one in time.

EXAMPLE 6.1

Lenssen et al. (2007a,b) reported the results of a multi-year study of the effect of crop rotation and tillage system on soil nitrate and soil water. Nine rotations under two tillage systems, conventional and no-till, were studied on a private farm near Havre, Montana. The experimental design was a split plot with repeated measures over depth and time. The whole plot treatments were crop rotations, and the split plot treatments were tillage systems. For each subplot, soil water and soil nitrate

doi:10.2134/2012.generalized-linear-mixed-models.c6

Analysis of Generalized Linear Mixed Models in the Agricultural and Natural Resources Sciences
Edward E. Gbur, Walter W. Stroup, Kevin S. McCarter, Susan Durham, Linda J. Young, Mary Christman, Mark West, and Matthew Kramer

were recorded at five depths. The study was conducted from 1998 to 2003. In this example, only soil nitrate in continuous spring wheat for the years 2000 to 2003 are considered. Because we will be considering only one rotation treatment in this example, the design reduces to a randomized complete block design with repeated measures over depth and time. The treatment was tillage system (*tillage*). Each block (*rep*) consisted of two plots, a conventional tillage plot and a no-till plot. For each plot, soil nitrates (*s_nitrate*) were measured at five depths (*depth*): 3, 9, 18, 30, and 42 inches. Hence, the effect of depth is a repeated measure over space. In addition, because the study was conducted over a series of years, and measurements were collected for each plot each year, the effect of year (*year*) is a repeated measure as well. Thus, this study has doubly repeated measures, namely, observations collected at different depths from the same plot and year and observations collected in different years from the same plot and depth.

An important point to note for the models we will be considering is that we expect the observations to be autocorrelated in space and time, even after adjusting for the main effects of year and depth; that is, the residuals (after fitting the main effects) are autocorrelated, and we need some way to model this autocorrelation. The reason that this occurs goes back to our expectation of the way things behave in time and space. Observations from two consecutive years or depths, even after adjusting for main effects, are still anticipated to be more similar to each other than to observations farther removed from each other. Thus, year and depth occur twice in the model, once as main effects, and then again when describing the correlation structure of the residuals. In the process of developing the model, we will test whether the anticipated correlation structure in time and space is present and, if not, we will return to an independent error structure.

SELECTION OF A DISTRIBUTION FOR SOIL NITRATE

The measurements of soil nitrates were non-negative, with an overall mean of 9.2 and a standard deviation of 8.8. The probability of obtaining a negative prediction for soil nitrate under the assumption of normality is substantial. The gamma and the lognormal distributions may be more appropriate models because they allow only positive values. As a first step, we decide on the distribution to use in the analysis and then turn to the correlation structure.

To gain insight into the data, we begin by analyzing the data for each year separately, assuming that the data are normally distributed and ignoring, for the moment, the correlation among depths from the same plot. The GLIMMIX statements are shown in Fig. 6–1. Note that the *random* statements could have been written as

random rep rep*tillage;

random depth / subject = rep*tillage residual;

FIG. 6-1. GLIMMIX statements to fit the normal distribution model in Example 6.1 for each year separately.

```
proc glimmix data=work plots=studentpanel;
   by year;
   class rep tillage depth;
   model s_nitrate = tillage depth tillage*depth / ddfm=kr dist=normal link=id;
   random intercept tillage / subject=rep;
   random _residual_ / subject=rep*tillage;
run;
```

In the second *random* statement, *depth* could have been omitted because the option *residual* clearly indicates that this is an R-side effect; i.e., the following statement would have given equivalent results:

*random/subject = rep*tillage residual;*

The alternative forms of these statements that are displayed in Fig. 6–1 are equivalent but more computationally efficient.

We anticipate that the soil nitrate measurements from the same plot will be correlated, even after adjusting for main effects, leading to a correlation structure among the residuals in the **R** matrix (sometimes called R-side correlation). In GLIMMIX the correlation among residuals is specified through the second *random* statement as shown in Fig. 6–1. The experimental unit on which the repeated measures were observed is specified using the *subject* option. Because no correlation structure was specified, the residuals are assumed to be independent. Thus, in this case, the results are the same whether or not this second *random* statement for the residuals is included.

Throughout the discussion, graphs for the 2002 data are presented; other years had comparable patterns in their graphs unless stated otherwise. In 2002 studentized residual plots under the assumption that the response is normally distributed with independent errors are displayed in Fig. 6–2. For all years (although not as obviously for 2002), the variability of the residuals was not constant over the range of predicted values. The remaining plots, especially the quantile (Q–Q) plots, further lead one to question the assumption of normality.

As alternatives to the normal distribution, both the gamma and the lognormal distributions were fit to these same data. The GLIMMIX statements are given in Fig. 6–3 and 6–4 for the gamma and lognormal distributions, respectively. The only change occurs in the *model* statement where the distribution and link function are specified.

The lognormal distribution tended to fit better than the gamma distribution for 2000 to 2002, but the gamma provided a slightly better fit in 2003. The studentized residual plots using the lognormal distribution with a log link for 2002 are displayed in Fig. 6–5. Because the lognormal tended to fit better for most years and did not fit badly in any year, this distribution was assumed for the remainder of the analysis.

FIG. 6-2. GLIMMIX output displaying plots of the conditional studentized residuals from the fit of the normal distribution model for the 2002 data in Example 6.1.

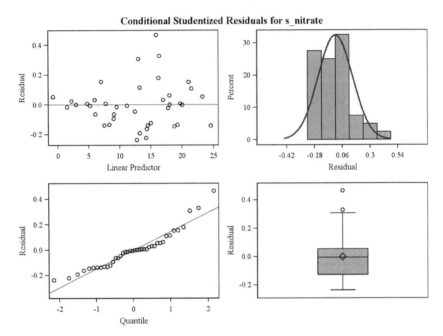

FIG. 6-3. GLIMMIX statements to fit the gamma distribution model in Example 6.1 for each year separately.

```
proc glimmix data=work plots=studentpanel;
   by year;
   class rep tillage depth;
   model s_nitrate = tillage depth tillage*depth / dist=gamma link=log ddfm=kr;
   random intercept tillage / subject =rep;
   random _residual_ / subject =rep*tillage;
run;
```

FIG. 6-4. GLIMMIX statements to fit the lognormal distribution model in Example 6.1 for each year separately.

```
proc glimmix data=work plots=studentpanel;
   by year;
   class rep tillage depth;
   model s_nitrate = tillage depth tillage*depth / dist=logn link=log ddfm=kr;
   random intercept tillage / subject =rep;
   random _residual_ / subject =rep*tillage;
run;
```

FIG. 6–5. GLIMMIX output displaying plots of the conditional studentized residuals from the fit of the lognormal distribution model in Example 6.1 for the 2002 data.

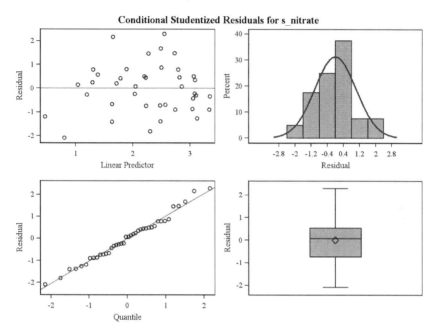

REPEATED MEASURES OVER DEPTH

Although independent errors have been assumed thus far, measurements from the same location at different depths would be expected to be correlated. To model the covariance structure, a first step is to estimate each variance and covariance parameter associated with the depth measurements, that is, to not impose any structure on the variances and covariances. This was accomplished by adding the *type = un* option in the second *random* statement in Fig. 6–6. A graph of the covariances between each depth (denoted by "Starting depth" in Fig. 6–7) and all deeper depths as a function of the distance between observations was constructed (Littell et al., 2006). Only the graph for the 2002 data is shown in Fig. 6–7; other years had a similar appearance. The unequal spacing of the depth measurements is clear from the graph.

For this study it seems reasonable to assume that the correlation between two observations would decrease as the distance between them increases until they become uncorrelated, so compound symmetry (Section 4.6), which has the same variance at each depth and the same covariance between all depths regardless of their distance from each other, is unlikely to model the data well. From Fig. 6–7, we see that values of the variances (i.e., covariances at a distance of zero) are small and possibly unequal. The largest covariances tend to be between neighboring depths, while covariances between distant depths are small or even negative. Thus, we consider a *un(2)* covariance structure (Section 4.5) that allows for differing variances at each depth and for different covariances between neighboring depths, but

FIG. 6-6. GLIMMIX statements to fit the lognormal distribution model with an unstructured covariance structure for depth in Example 6.1 for each year separately.

```
proc glimmix data=work plots=studentpanel;
  by year;
  class rep tillage depth;
  model s_nitrate = tillage depth tillage*depth / dist=logn link=log ddfm=kr;
  random intercept tillage / subject =rep;
  random depth / residual type=un vcorr subject =rep*tillage;
run;
```

FIG. 6-7. Plot of the variances and covariances of soil nitrate between depths as a function of distance for the lognormal distribution model with an unstructured covariance structure for depth in Example 6.1 for the 2002 data.

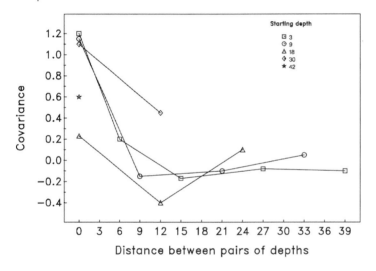

Distance between pairs of depths

sets covariances for non-neighboring depths equal to zero. The GLIMMIX statements for this model of the covariance structure remain unchanged except for changing the *type* to *un(2)* in the second *random* statement in Fig. 6–6.

If the depths had been equally spaced, a first order autoregressive correlation structure (*ar(1)*) might be anticipated (Section 4.6). This correlation structure agrees with our intuition that the correlation should decrease as observations become further apart. However, because of the unequal spacing of the observations, *ar(1)* may be inappropriate, and we should consider other potential covariance structures.

When observations are unequally spaced, spatial covariance functions often are useful for modeling the covariance structure (even for repeated measures in time). To use these functions, depth must be a continuous numerical variable, not a class variable. A programming statement in a *data* step can be used to define a new variable *d2* to be equal to *depth*. Although the variables *depth* and *d2* are numerically equal, one (*depth*) is a classification variable and the other (*d2*) is a continuous numerical variable. This allows us to specify the fixed effect of depth as a class variable in the *model* statement and the random effect of depth as a continuous variable to estimate the semivariogram. Both the power and exponential covariance structures were considered using *type* = *sp(pow)(d2)* and *type* = *sp(exp)* (*d2*), respectively, on the second *random* statement in Fig. 6–6.

The conditional studentized residuals and the AICCs were used to compare fits among the different covariance structures. When the lognormal distribution is specified, GLIMMIX computes the logarithm of the response (nitrates) and then analyzes the transformed variable assuming the normal distribution. In this case, no pseudo-data are generated when fitting the model, regardless of the method used. Thus, because we are fitting the normal distribution and only the covariance structure changes, the use of AICC is appropriate here. However, if we had decided to use the gamma distribution with a log link function to model the distribution of the response, the AICC comparison would not have been appropriate.

In Table 6–1 the AICCs for the unstructured covariance (*un*) and for the *un(2)*, where only neighboring depths had non-zero covariances, were larger than the other covariance structures for all years, indicating a poorer fit. With the exception of 2002, compound symmetry did not provide as good of a model fit as the other covariance structures. The model assuming independence in the depth measurements performed well in 2002 and 2003, but not as well in 2000 or 2001. The first order autoregressive, power spatial, and exponential spatial covariance models all performed similarly, with the exponential spatial covariance structure performing slightly less well in 2003. There is little distinction between the power spatial covariance structure and the first order autoregressive based on their AICCs. Their AICC values were within one unit of each other for all years, a difference that is

TABLE 6–1. Comparison of AICC values for several R-side covariance structures for the lognormal distribution model in Example 6.1 for each year separately.

Covariance structure	GLIMMIX type	AICC			
		2000	2001	2002	2003
Independence	–	88.38	59.39	64.66	89.25
Compound symmetry	*cs*	91.06	62.06	65.09	91.93
First order autoregressive	*ar(1)*	85.70	55.44	67.01	87.84
Spatial, exponential covariance function	*sp(exp)*	84.69	54.81	66.62	91.93
Spatial, power covariance function	*sp(pow)*	84.69	54.81	66.62	87.95
Unstructured	*un*	148.31	125.39	149.08	153.04
Unstructured, covariance = 0 for non-adjacent depths	*un(2)*	100.57	76.25	85.84	106.63

FIG. 6-8. GLIMMIX output displaying plots of the conditional studentized residuals as a function of distance treated as a continuous variable for the lognormal distribution model with a power spatial covariance structure in Example 6.1 for the 2002 data.

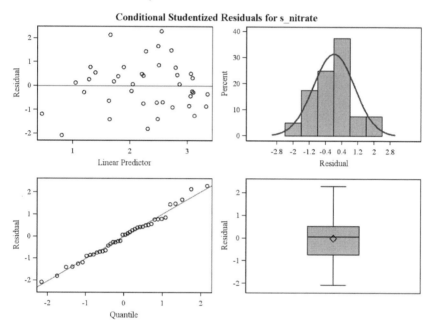

unlikely to impact the results. When comparing the studentized residual plots, the power spatial covariance structure was a little better (Fig. 6–8).

Because the independence covariance structure produced AICC values similar to the $ar(1)$ structure and spatial covariance function structures, a natural question is whether it is really necessary to model the covariance structure. Is an assumption of independence in the error structure (the **R** matrix) appropriate here? The assumption may be tested in GLIMMIX by adding a *covtest* statement with the test specification *cindep*. The conditional test of independence (*cindep*) compares the covariance structure specified for the **R** matrix to that when **R** is restricted to being diagonal (no non-zero covariances), without modifying the G-side structure. In contrast, the test specification *indep* tests the null hypothesis that **R** is diagonal and that there are no G-side effects. For 2000, 2001, 2002, and 2003, the p-values associated with the null hypothesis of independence against the first-order autoregressive alternative were 0.0546, 0.0324, 0.4694, and 0.2538, respectively. Thus, modeling the covariance structure is important in the first 2 years but not the latter 2 years. Notice that this is consistent with the conclusions drawn from comparisons of the AICC values.

When the variances differ substantially from each other, it can be difficult to separate the strength of the correlation from the heterogeneity of the variances when modeling the covariance structure. In this case, using the covariance estimates to consider heterogeneity of variances and then turning to the estimated

TABLE 6-2. *p*-values for tests of significant interaction between tillage system and depth for several combinations of distribution and covariance structure in Example 6.1.

		p-value for Testing Tillage System × Depth Interaction			
Distribution	Covariance structure	2000	2001	2002	2003
Normal	Independence	0.8679	0.0129	0.1382	0.6947
Gamma	Independence	0.7515	0.2485	0.0071	0.1585
Lognormal	Independence	0.5741	0.2728	0.0142	0.3391
Lognormal	Compound symmetry	0.5741	0.2728	0.0243	0.3391
Lognormal	First order autoregressive	0.5967	0.1011	0.0235	0.3297
Lognormal	Spatial (power)	0.5334	0.1475	0.0263	0.3322
Lognormal	Spatial (exponential)	0.4604	0.0769	0.0287	0.4162
Lognormal	Unstructured (2)	0.2869	0.1973	0.5753	0.6899

correlations to gain insight into correlation structure may be a better approach. The *vcorr* option requests that the correlation matrix for the residuals be printed.

Of course, primary interest lies not in the specific form of the distribution or in the covariance structure, but in the tests of the fixed effects. The *p*-values associated with the test of the null hypothesis of no tillage system × depth interaction are given in Table 6–2 for several combinations of distribution and covariance structure.

The difference between the normal distribution and the other distributions is most notable for 2001 and 2002. Assuming a normal distribution and independent errors, one would conclude that tillage system and depth interacted in 2001 but not in 2002; the opposite conclusion would have been drawn using either the lognormal or the gamma distribution. This clearly illustrates the importance of developing a model that adequately fits the data. However, notice that for the three covariance structures that fit best (first order autoregressive, spatial power, and spatial exponential), the *p*-values were close and the inference the same for all 4 years. In general, it is important to get a reasonable model of the covariance structure, but more than one covariance structure may provide an adequate fit. Thus, choosing any of these three covariance structures would be acceptable.

We chose to use a lognormal distribution with a log link function and a first order autoregressive covariance structure. The reason for this choice is that it makes sense to use a consistent covariance structure throughout an analysis. As we will see when we consider the full analysis, the spatial covariance functions are not options in doubly repeated measures, but the first order autoregressive is. Thus, given little to choose from among the three, we chose the structure that we could carry forward into the full analysis.

DOUBLY REPEATED MEASURES

Having explored the data for each year separately, an analysis of the entire dataset will be conducted. Two repeated measures are present: depth and year. If year is considered to be a random effect, then the correlation among years is modeled in the random effects covariance matrix **G**, and the correlation among depths is modeled in the residual covariance matrix **R**. However, because the effect of planting spring wheat continuously could depend on the number of years that the soil has been planted to continuous spring wheat, year was assumed to be a fixed effect. Thus, two fixed effects, depth and year, have been observed repeatedly, and this needs to be reflected in our analysis.

Galecki (1994) first proposed using Kronecker products as a means to structure the covariance matrix when there are repeated measures for two fixed effects. To conduct the analysis in SAS (SAS Inst., Cary, NC) without the use of programming statements, an unstructured covariance (*un*) must be used for one of the two effects. The other may also be unstructured, compound symmetric (*cs*) or first order autoregressive (*ar(1)*). Further, GLIMMIX does not currently offer this option. So, we must either develop the programming statements ourselves or use PROC MIXED. Using MIXED requires that we must assume normality of the observed values or transform the values so that they are approximately normal. Fortunately the lognormal distribution is derived by assuming that the natural logarithm of the observed values is normally distributed. Thus, we can simply analyze the logarithm of the response (*l_s_nitrate*) and back transform the estimated means and standard errors, the latter using the delta method. The MIXED statements to fit the model with unstructured covariance matrices for both year and depth are given in Fig. 6–9. Because *year* is listed before *depth* in the *repeated* statement, the covariance structure listed before the @ symbol is associated with year and the structure after the @ symbol is associated with depth.

Various combinations of covariance structures were fit. In Table 6–3, each candidate covariance structure was compared to the *un@un* covariance structure using a likelihood ratio test (Roy and Khattree, 2005). Using *un* for the year covariance structure, neither *un@cs* nor *un@ar(1)* was significantly different in fit from the *un@un*. By reversing the order of *year* and *depth* in the *repeated* statement, the unstructured covariance structure is applied to depth, and we can see that a

FIG. 6–9. MIXED statements for the fit of a normal distribution to the natural logarithm of soil nitrate with unstructured covariance structures for both year and depth in Example 6.1.

```
proc mixed data=work covtest;
    class rep tillage depth year;
    model l_s_nitrate = tillage depth tillage*depth year year*tillage year*depth
        year*tillage*depth / ddfm=kr;
    random intercept tillage / subject=rep;
    repeated year depth / type=un@un subject=rep*tillage;
run;
```

TABLE 6-3. Comparison of covariance structures for doubly repeated measures (year and depth) for the lognormal distribution model in Example 6.1.

Year Covariance	Depth covariance	Number of parameters	−2log(L)	Comparison to un@un		
				Difference in −2log L	Difference in df	p-value
un	un	25	242.3			
un	cs	11	260.9	18.6	14	0.1808
un	ar(1)	11	253.5	11.2	14	0.6703
cs	un	16	257.3	15.0	9	0.0909
ar(1)	un	16	259.5	17.2	9	0.0457

FIG. 6-10. MIXED statements for the fit of a normal distribution to the natural logarithm of soil nitrate with an unstructured covariance structure for year and a first order autoregressive covariance structure for depth in Example 6.1.

```
proc mixed data=work covtest;
    class rep tillage depth year;
    model l_s_nitrate = tillage depth tillage*depth year year*tillage year*depth
        year*tillage*depth / dist=normal link=id ddfm=kr;
    random intercept tillage / subject=rep;
    repeated year depth / type=un@ar(1) subject=rep*tillage;
    lsmeans year*depth;
    ods output lsmeans=lsm;
run;
```

simpler covariance structure for year (i.e., *cs* or *ar(1)*) did not fit as well as *un@un*. Based on our earlier work with depth and the slightly greater likelihood function, we would choose to use the *un@ar(1)* structure. The MIXED statements to fit this model are shown in Fig. 6–10.

Once the proper distribution and covariance structure has been determined, the significance of the fixed effects can be evaluated (Fig. 6–11). For continuous spring wheat, tillage had no significant effect on soil nitrates, but depth and year interacted significantly. To understand this interaction, a plot of the least squares means is generally helpful. Unlike in GLIMMIX, these plots are not readily available in MIXED but can be created. A data set containing the least squares means must be created using an *ods output* statement. Then that data set is used to create the plot shown in Fig. 6–12. It becomes evident that the interaction was largely due to the differences in the first 2 years (2000 and 2001) and the last 2 years (2002 and 2003). For the first 2 years, soil nitrate dropped when going from 9 to 18 inches but increased from 9 to 18 inches in the last 2 years. ∎

FIG. 6-11. MIXED output containing tests of the fixed effects in the lognormal distribution model with an unstructured covariance structure for year and a first order autoregressive covariance structure for depth in Example 6.1.

Type III Tests of Fixed Effects

Effect	Num DF	Den DF	F Value	Pr > F
tillage	1	10.3	0.56	0.4701
depth	4	11.8	6.93	0.0041
tillage*depth	4	11.8	0.63	0.6513
year	3	17.6	9.10	0.0007
tillage*year	3	17.6	1.73	0.1970
depth*year	12	25.9	8.96	<0.0001
tillage*depth*year	12	25.9	1.76	0.1111

FIG. 6-12. Least square means of soil nitrates for the year × depth interaction in the lognormal model with an unstructured covariance structure for year and a first order autoregressive covariance structure for depth in Example 6.1.

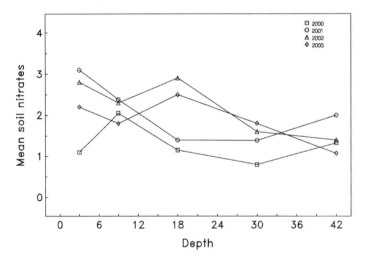

6.3 ANALYSIS OF A PRECISION AGRICULTURE EXPERIMENT

Statistical analysis of precision agriculture experiments offers a number of challenges not commonly found in traditional field experiments. In addition to datasets routinely containing thousands of observations having variables obtained from a variety of sources with varying measures of reliability, successfully accounting for spatial correlation in a mixed model can be a non-trivial task even for experienced researchers and their statistician. The following example illustrates some of the issues involved.

EXAMPLE 6.2

The data for this example were obtained from an on-farm field trial on a commercial cotton farm in northeast Louisiana (Burris et al., 2010) and are used with permission. The trial was conducted on a 33-acre field with a history of root-knot nematode problems. Nematodes cause damage to the root system of the cotton plant that can severely inhibit the uptake of nitrogen. The purpose of the study was to evaluate the effects of a nematicide and three nitrogen rates on cotton lint yield to develop a treatment prescription for future use on that field. In this example we focus on a statistical analysis that can serve as the basis for developing such a treatment prescription.

DESCRIPTION OF THE FIELD TRIAL

From prior research the field used in the experiment was known to vary spatially with respect to soil type. Apparent soil electroconductivity (EC_a) has been shown to correlate well with soil clay content and was used as a proxy for soil type. EC_a measurements were taken across the entire field. From the raw EC_a data an ordinal variable defining three soil-type categories representing low, medium, and high clay content was created. The intention of the researcher was to use these three EC_a zones as management zones in developing treatment prescriptions for the nematicide and the nitrogen fertilizer. Figure 6–13 shows a map of the field with the three soil categories.

FIG. 6–13. Plot of the three ECa zones in the field and the field's true orientation with respect to easting and northing coordinates in Example 6.2. The 24-row-wide plots are also shown.

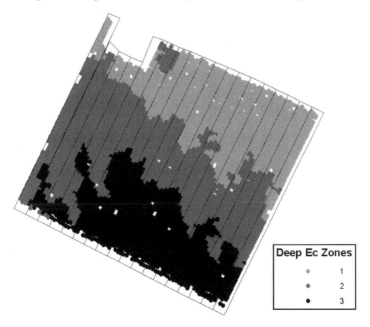

Deep Ec Zones

※	1
●	2
●	3

FIG. 6-14. Relationships among the passes within a 24 row plot defined by a nitrogen rate–nematicide usage treatment combination in Example 6.2.

Structure of 24-Row Plots						
Nitrogen Application Passes	Nitrogen Pass 1 (12 rows)			Nitrogen Pass 2 (12 rows)		
Nematicide Application Passes	Nematicide Pass 1 (4 rows)	Nematicide Pass 2 (4 rows)	Nematicide Pass 3 (4 rows)	Nematicide Pass 1 (4 rows)	Nematicide Pass 2 (4 rows)	Nematicide Pass 3 (4 rows)
Field Rows	1 2 3 4 5 6	7 8 9 10 11 12	13 14 15 16 17 18	19 20 21 22 23 24		
Harvest Passes	Harvest Pass 1 (6 Rows)	Harvest Pass 2 (6 Rows)	Harvest Pass 1 (6 Rows)	Harvest Pass 2 (6 Rows)		

Two factors were used in the experiment: an application of a nematicide and of nitrogen fertilizer. The nematicide treatment consisted of two levels, either not applied or applied at a fixed rate of 3 gallons per acre. The three nitrogen rates used were 80, 115, and 135 pounds per acre.

The experiment was laid out in three replicates, with the six nitrogen–nematicide treatment combinations assigned at random to plots within each replication. These plots, oriented from the southwest to northeast, extended the length of the field and were each 24 rows of cotton plants wide. Nitrogen application equipment spanned 12 rows, requiring two nitrogen application passes within each treatment plot. Hence, nitrogen application passes were nested within the 24-row-wide treatment plots. The nematicide application equipment spanned four rows, requiring three nematicide application passes per nitrogen application pass. Thus, nematicide application passes were nested within nitrogen application pass.

At harvest, a yield monitor on the cotton picker measured cotton lint yield every 2 seconds as it traversed the field. Yield data were spatially referenced using a GPS receiver mounted on the picker. The cotton picker spanned six rows. Harvest passes were nested within nitrogen application pass with two harvest passes per nitrogen application pass. Note that the two harvest passes within a nitrogen application pass each covered half of the middle nematicide application pass within that nitrogen application pass. Figure 6–14 shows the relationships among the various passes.

Yield data were loaded into GIS software, cleaned, and then scaled to pounds per acre. The cotton lint yield and field characteristic data for each sampled yield location (i.e., each 2-second lint collection) were combined into a single data file with one line in the file per sampled location. The data file contained 6008 yield observations, a relatively small dataset for this type of application.

PRELIMINARY DESCRIPTIVE STATISTICS FOR LINT YIELD

Table 6–4 presents a two-way breakdown of observed mean cotton lint yield by EC_a zone and nitrogen rate. The observed mean lint yield for the entire field was

TABLE 6-4. Observed mean lint yield by EC_a zone and nitrogen rate in Example 6.2.

EC$_a$ zone	Nitrogen rate			EC$_a$ zone means
	80	115	135	
1–High clay	1293.11	1292.98	1291.74	1292.60
2–Medium clay	1339.16	1360.83	1311.65	1338.32
3–Low clay	1310.84	1314.49	1293.85	1306.78
Nitrogen rate means	1316.70	1329.49	1300.43	1315.74

TABLE 6-5. Observed mean lint yield by EC_a zone and nematicide usage in Example 6.2.

EC$_a$ zone	Nematicide usage		EC$_a$ zone means
	Not applied	Applied	
1–High clay	1268.81	1312.89	1292.60
2–Medium clay	1318.36	1362.95	1338.32
3–Low Clay	1298.64	1314.20	1306.78
Nematicide usage means	1299.55	1332.16	1315.74

1315.74 pounds per acre. Among the three EC_a zones overall, zone 2 (medium clay content) had the highest observed mean yield, followed by zone 3 (low clay content), and finally zone 1 (high clay content). Since nematodes tend to be more prevalent in soils with high clay content, the fact that zone 1 has the lowest observed mean yield is not unexpected. Broken down by nitrogen application rate, the rate of 115 pounds per acre had the highest overall observed mean yield, followed by the 80 pounds per acre rate. The highest rate of nitrogen application, 135 pounds per acre, had the lowest observed mean cotton lint yield.

The overall ranking of nitrogen rates described above is preserved in both EC_a zones 2 and 3. However, in zone 1 the 80 pounds per acre rate had the highest observed mean yield, followed by the 115 pounds per acre rate, and finally the 135 pounds per acre rate, although these zone 1 means are only slightly different across the three nitrogen rates. This similarity in yield means is not unexpected, due to the higher prevalence of nematodes in this zone and the damage they do to the roots of the cotton plant. This difference in trends across the three EC_a zones suggests the possibility of a two-way interaction between EC_a zone and nitrogen rate.

Table 6–5 presents a two-way breakdown of observed mean cotton lint yield by EC_a zone and nematicide usage. Overall, the areas in which the nematicide was used had an observed mean lint yield that was about 32 pounds per acre higher

TABLE 6-6. Observed mean lint yield by EC_a zone, nitrogen rate and nematicide usage in Example 6.2.

EC$_a$ zone	Nitrogen rate	Nematicide usage	
		Not applied	Applied
1–High clay	80	1262.43	1319.54
	115	1259.24	1325.02
	135	1285.22	1296.74
2–Medium clay	80	1332.65	1349.56
	115	1338.27	1282.42
	135	1281.94	1349.32
3–Low clay	80	1311.76	1310.29
	115	1303.07	1327.24
	135	1283.02	1306.02

than in those areas where it was not used. Within both EC_a zones 1 and 2, the areas in which the nematicide was used had observed mean yields that were 44 pounds per acre higher than those areas where the nematicide was not applied. On the other hand, in zone 3 the areas in which the nematicide was applied had an observed mean yield that was only about 16 pounds per acre higher than those areas where it was not. Again, this is as expected since nematodes are more likely to be present in EC_a zones 1 and 2 than they are in zone 3. The differences in the effects of nematicide usage suggest the existence of a two-way interaction between EC_a zone and nematicide usage.

Table 6–6 gives a three-way breakdown of observed mean cotton lint yield by EC_a zone, nitrogen rate, and nematicide usage. Inspection of the values provided by this breakdown suggests the possibility of a three-way interaction among these factors.

These initial assessments of the two- and three-way data summaries suggest the possibility of interactions between the variable used to define field management zones (EC_a zone) and the applied nitrogen and nematicide treatments. Hence, a variable rate treatment prescription may be appropriate for one or both of the applied treatments. Formal assessment of the significance of these interactions will be based on the statistical models discussed below.

INITIAL MIXED MODEL ANALYSIS OF COVARIANCE

The measured response variable, cotton lint yield, contains several sources of variability that can be divided into the following categories: the applied treatments, the observed field characteristics, the unobserved field characteristics, and the variability induced by the way the experiment was conducted.

The applied treatments consist of nitrogen rate (*nrate*) and nematicide (*nem*). The observed field characteristic considered in this analysis is the EC_a zone (*ec_zone*), which, as described previously, serves as a proxy for soil type and which ultimately is to be used to define field management zones. A mixed model analysis of covariance was used to model cotton lint yield (*yld*) as a function of the applied treatments and the measured field characteristic. Yield was assumed to be normally distributed. GLIMMIX was used to fit this model. The variables *nrate*, *nem*, and *ec_zone* were fixed effects. Main effects and all two- and three-way interactions between these factors were included in the model. In addition, spatial coordinate variables were included in the model as continuous covariates to account for a possible linear trend across the length and breadth of the field. These variables are labeled *loc_x* and *loc_y*, and represent rescaled versions of the *easting* and *northing* spatial coordinate variables used to identify the spatial location of each yield point.

There are several sources of variation in the response variable resulting from the way the experiment was conducted that should be included in the model as random effects. Because the treatments were randomized to 24-row-wide plots (*plot*) within the three replications (*rep*) separately, *rep* and *rep*plot* were included in the model as random effects. In addition, there is potential variation among the nitrogen application passes (*apass*) and the cotton picker harvest passes (*hpass*); hence, these variables were included as random effects as well. There is also variability in the response due to the nematicide application passes. However, it is not possible to include random effects for the nematicide application passes. The reason becomes clear on inspection of Fig. 6–14. Within a given nitrogen application pass, random effects for the two outer-most nematicide passes would completely coincide with the random effects for the harvest passes in which they are contained. The random effect for the middle nematicide application pass that is shared by the two harvest passes would completely coincide with the random effect for the nitrogen application pass itself. Hence if random effects for the nematicide application passes were included, the result would be an unidentifiable model (i.e., the parameters would not be identifiable). The variability due to nematicide application pass is not lost or ignored, however. The random effect *hpass* actually accounts for the combined variability of the harvest pass and the nematicide application pass that is completely contained within that harvest pass. The random effect *apass* accounts for the combined variability of the nitrogen application pass and the middle nematicide application pass shared by the two harvest passes it contains.

We first consider a model that assumes that the model errors (i.e., the R-side random effects) are independent and identically distributed normal variates with constant variance. The GLIMMIX statements to fit the model are shown in Fig. 6–15. The GLIMMIX output in Fig. 6–16 shows the dimensions of the **X** and **Z** design matrices and the number of covariance parameters in the model. There were three *reps* and 18 (24-row) *plots* used in the experiment. In addition, there were 36 nitrogen application passes. One of the application passes contained a single harvest pass, while the rest contained two. Hence, there are $3 + 18 + 36 + (35 \times 2) + 1 = 128$ random effects in the model, as reflected in the number of columns

FIG. 6-15. GLIMMIX statements to fit the initial model in Example 6.2.

```
proc glimmix data=yield plots=(studentpanel(conditional));
  class ec_zone nrate nem rep plot apass hpass ;
  model yld = ec_zone | nrate | nem  loc_x loc_y / dist=normal link=id
    ddfm=satterth;
  random rep;
  random rep*plot;
  random rep*plot*apass;
  random rep*plot*apass*hpass;
  nloptions gconv=0;
run;
```

FIG. 6-16. GLIMMIX output containing the dimension information from the initial model in Example 6.2.

Dimensions	
G-side Cov. Parameters	4
R-side Cov. Parameters	1
Columns in X	50
Columns in Z	128
Subjects (Blocks in V)	1
Max Obs per Subject	6008

in Z shown in the table. There are four G-side covariance parameters in the model, corresponding to the variances of the random effects associated with *rep*, *plot*, *apass*, and *hpass*. The single R-side covariance parameter is the variance of the error term. The field is the sole subject with 6008 observations. We point out that precision agriculture datasets can produce much larger datasets than this one. This is actually one of the smallest precision agriculture datasets we have analyzed.

The covariance parameter estimates are given in Fig. 6–17. Note that although *rep* was initially conceived of as a blocking factor when the researchers designed the study, the field locations corresponding to the *reps* were very large, and there was apparently little variability among these three sections of the field. As a result the variance of the *rep* random effect is estimated to be zero. Because this variance estimate is on the boundary of the parameter space, the standard error is set to missing by GLIMMIX. The rest of the variance component estimates are numerically greater than zero, although not all appear to be significantly greater than zero. The variability among *plots* appears to be marginally significant. There does not appear to be significant variability among nitrogen application passes. On the other hand, there is significant variability among harvest passes. There is also a

FIG. 6-17. GLIMMIX output containing estimates of the covariance parameters for the initial model in Example 6.2.

Covariance Parameter Estimates		
Cov Parm	Estimate	Standard Error
rep	0	.
rep*plot	842.41	557.72
rep*plot*apass	96.3362	305.37
rep*plot*apass*hpass	1161.69	319.22
residual	15649	287.53

FIG. 6-18. GLIMMIX output containing the tests of the fixed effects for the initial model in Example 6.2.

Type III Tests of Fixed Effects				
Effect	Num DF	Den DF	F Value	Pr > F
ec_zone	2	5947	165.21	<0.0001
nrate	2	10.59	0.47	0.6389
ec_zone*nrate	4	5925	4.62	0.0010
nem	1	10.54	3.27	0.0990
ec_zone*nem	2	5913	9.19	0.0001
nrate*nem	2	10.56	0.06	0.9409
ec_zone*nrate*nem	4	5927	2.14	0.0727
loc_x	1	10.96	9.57	0.0103
loc_y	1	5952	23.31	<0.0001

great deal of residual variation in the data, as evidenced by the large estimate of the residual variance as compared to the variances of the other random effects.

Figure 6–18 gives the results of the F-tests for the fixed effects. The three-way interaction between *ec_zone*, *nrate*, and *nem* is not significant at the 0.05 level, but it is significant at the 0.10 level. The two-way interaction between *ec_zone* and *nrate* and the two-way interaction between *ec_zone* and *nem* are both highly significant. This implies that the effects of *nrate* and *nem* depend on the *ec_zone* to which they are applied, and, hence, that development of a variable-rate treatment prescription involving *nrate* and *nem* would be appropriate for the field. The levels of *ec_zone* comprise the field management zones within which the various *nrate* × *nem* treatment combinations would be compared in developing a treatment prescription for the field.

Before we proceed, we should assess the model and check for violations of the model assumptions. Figure 6–19 contains graphs produced by GLIMMIX for the conditional studentized residuals that are useful for checking the normality and homogeneity of variance assumptions of the model errors. The empirical

FIG. 6-19. GLIMMIX output displaying plots of the conditional studentized residuals for the initial model in Example 6.2.

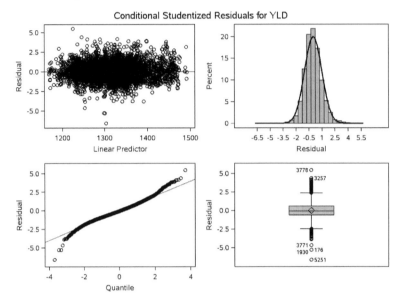

distribution of the residuals is fairly symmetric. From the quantile plot, the tails of the distribution appear to be somewhat heavier than that of a normal distribution, but otherwise, the normality assumption does not appear to be violated to any great extent.

In addition to the normality assumption, this model assumes that the errors are independent. If the conditional studentized residuals from the fitted model exhibit spatially correlation, this would indicate a possible violation of that independence assumption. The independence assumption can be checked by assessing the spatial correlation among the conditional studentized residuals. Spatial correlation can be described in terms of spatial variation. If residuals tend to be more alike when they are from locations close together than when they are from locations farther apart, the residuals exhibit spatial correlation, with residuals close together being more highly correlated than residuals farther apart. Hence, assessing the spatial correlation among the residuals can be accomplished by assessing the spatial variation of the residuals. The semivariogram (Cressie, 1993; Schabenberger and Gotway, 2005) is a tool that is used to assess the level and extent of the spatial variation that exists among the studentized residuals from a model. Theoretical semivariograms are non-decreasing functions of distance that give a measure of the spatial variability between residuals separated by a given distance. Their graphs are either flat, or rise initially and then level out. The height on the graph at which the semivariogram levels out corresponds to the residual variance, and the distance at which the semivariogram reaches this height is called the range. Residuals separated by a distance greater than the range are uncorrelated, while residuals within that distance are spatially correlated. A flat semivariogram indicates constant spatial variation and hence zero spatial correlation. A non-constant semivariogram that rises initially and then levels out indicates the existence of non-zero spatial correlation at closer distances that becomes attenuated as the distance between residuals increases.

The VARIOGRAM procedure in SAS was used to estimate the semivariogram function for the conditional residuals of the model described above. A graph of the estimated semivariogram is given in Fig. 6–20. It clearly shows that residuals close together are more similar than residuals farther apart. It has been constructed so that the first lag class, called lag class zero, has a width of 5 distance units and hence is measuring spatial variation for residuals that are anywhere from 0 to 5 distance units apart. The other lag classes are each 10 distance units in width. Lag class one includes distances from 5 to 15 distance units, lag class two includes distances from 15 to 25 distance units, and so on. Distance units are based on northing and easting coordinates.

The results of this variogram analysis can be expressed in terms of spatial correlation. In particular, the correlation between residuals separated by distances contained in lag class zero (0–5 distance units) can be shown to be approximately 0.417. This is considered a moderate to strong level of correlation. If one residual was used to predict another within this distance class using simple linear regression, such a model would account for 17.4% of the variability in the residual being predicted; i.e., $R^2 = 0.174$. While not great from a prediction standpoint, it is not a negligible amount if we would prefer it to have been zero. The correlation between residuals separated by distances contained in lag class one (5–15 distance units) is

FIG. 6-20. Plot of the empirical semivariogram of the conditional studentized residuals for the initial model in Example 6.2.

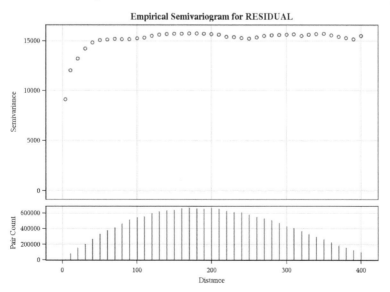

Empirical Semivariogram for RESIDUAL

220 CHAPTER 6

estimated to be 0.231. This is considered a negligible (or at most a weak) level of correlation. A simple linear regression model using one residual to predict another within this distance class would account for only about 5.3% of the variability in the residual being predicted ($R^2 = 0.053$). The correlations between residuals separated by greater distances are negligible as well. For lag class two (15–25 distance units), the estimated correlation is 0.154 ($R^2 = 0.024$); for lag class three (25–35 distance units) it is 0.090 ($R^2 = 0.008$); for lag class four (35–45 distance units) it is 0.052 ($R^2 = 0.003$); and for lag class five (45–55 distance units) it is 0.037 ($R^2 = 0.001$).

Adjacent locations within a harvest pass are approximately 4.2 distance units apart. Adjacent harvest passes are separated by approximately 5.7 distance units. Therefore lag class zero can be interpreted as measuring the spatial variation of residuals from adjacent locations within the same harvest pass. Lag class one and beyond are measuring the spatial variation between residuals in non-adjacent locations in the same harvest pass and also between residuals in different harvest paths. Based on the results above we see that adjacent residuals in the same harvest pass are moderately to strongly correlated, while non-adjacent residuals in the same harvest pass and residuals in different harvest passes are at most weakly correlated. This suggests a violation of the assumption of independent errors imposed on the data by this model. Caution should therefore be exercised in drawing conclusions from this model.

INCLUSION OF SPATIAL VARIATION EFFECTS IN THE MODEL

If the error terms are in fact not independent, then the model considered above is too restrictive. This problem can be addressed in several ways. We could consider fitting a more general model that allows the error terms to be correlated. The GLIMMIX procedure includes several spatial covariance error structures that can be used for this purpose. To this end, we first attempt to fit a model having a spherical spatial covariance error structure. The GLIMMIX statements are shown in Fig. 6–21. This covariance structure is incorporated into the model through the *type* option on the fifth *random* statement. The keyword *_residual_* instructs GLIMMIX

FIG. 6–21. GLIMMIX statements to fit the model using a spherical spatial error covariance structure in Example 6.2.

```
proc glimmix data=yield  plots=(studentpanel(conditional));
    class ec_zone nrate nem rep plot apass hpass ;
    model yld = ec_zone | nrate | nem loc_x loc_y / dist=normal link=id
        ddfm=satterth;
    random rep;
    random rep*plot;
    random rep*plot*apass;
    random rep*plot*apass*hpass;
    random _residual_ / type=sp(sph)(easting northing) ;
run;
```

that the following specifications apply to the R-side covariance structure. The variables *easting* and *northing* are the spatial coordinates of the yield measurements and are used to compute the distance between pairs of points. Unfortunately, with 6008 observations the memory required by GLIMMIX to fit this spatial structure exceeded the 2 GB of memory available to SAS on our computer, and the program aborted without completing the analysis. GLIMMIX also ran out of memory when trying to fit a spatial exponential structure. This problem results from the combined effects of the following four factors:

1. the number of observations in the dataset,
2. the amount of memory available to SAS, which in turn is limited by the amount of memory installed on the computer,
3. the type of model being fit to the data, and
4. limitations imposed by the implementation of the software being used to fit that model.

For a given dataset, the number of observations is fixed, so we do not have control over issue 1. We could address issue 2 by adding more memory to our computer. However, it is not uncommon for precision agriculture datasets to be much larger than the one considered here. In addition, we expect precision agriculture experiments to become larger and more complex in the future so that, regardless of how much additional memory we might install now, the problem would eventually reappear. One could attempt to address item 4 by searching for software with a more efficient implementation, perhaps designed specifically for analyzing large precision agriculture datasets of this type. However, even if such specialized software exists, there would be costs associated with its procurement and/or learning how to use it. Our solution in the remainder of this example will be to address issue 3 by using a different type of model, one that requires fewer computer resources to fit.

Spatial correlation can result from the effects of unmeasured or otherwise unaccounted for field characteristics that vary spatially across a field. Failure to adequately account for spatial trend in a precision agriculture dataset can induce spatial correlation among model residuals (McCarter and Burris, 2010). Residual spatial correlation may reflect unaccounted for trends. To the extent that such trends can be accounted for by including additional explanatory variables, the residual spatial correlation should be reduced or even eliminated. If georeferenced field characteristics have been measured across a field, then those characteristics can be used in a parametric model in an attempt to account for observed spatial trend. The available field characteristics may or may not be successful in accounting for all existing trends.

What can be done if the available measured field characteristics do not adequately account for the observed spatial trend in the data? Nonparametric smoothing splines can be used to account for residual trend in the data. They are very flexible and are incorporated in a model using spatial coordinates that will be available when observations are georeferenced. When a nonparametric smoothing spline is combined with parametric model components, the resulting model is called a semi-parametric model. Certain types of semi-parametric models containing penalized smoothing splines have representations as mixed models, and therefore can be fit with mixed-model software (Ruppert et al., 2003). The GLIMMIX procedure has the ability to fit such semi-parametric models through the inclusion of a penalized radial smoothing spline that is incorporated in a model as a random effect. Our goal in using such a semi-parametric model is to improve the validity of inferences involving the treatments in a precision agriculture experiment. This is accomplished by using a nonparametric smoothing spline to account for residual spatial trend, which in turn can reduce, and potentially eliminate, residual spatial correlation. If the radial smoother is able to account for enough of the spatial variation so that the resulting spatial correlation is negligible, then a model that assumes an independent error structure, which uses fewer computing resources, can be used.

Our initial attempt at using a radial smoother is given by the GLIMMIX statements in Fig. 6–22. The format of the *random* statement differs from that used in Fig. 6–21. The *type = rsmooth* option adds a nonparametric radial smoother to the parametric model already considered, resulting in a semi-parametric model. As before, the variables *easting* and *northing* allow GLIMMIX to compute the distance between points. To fit a radial smoother, a number of knots must be placed across the field. The *knotmethod* option specifies the method by which GLIMMIX selects and positions the knots used by the radial smoother. By default, the knot locations are automatically selected using a kd-tree (in this case, a $k = 2$ dimensional tree). A kd-tree is a data structure that can be used to partition a k-dimensional dataset into subsets of roughly equal size (called the bucket size). The vertices of

FIG. 6–22. GLIMMIX statements to fit the model with a radial smoother with knots selected using a kd-tree in Example 6.2. A plot of the location of the knots is also produced.

```
ods output  knotinfo=knotinfo;
proc glimmix data=yield ;
    class ec_zone nrate nem rep plot apass hpass ;
    model yld = ec_zone | nrate | nem easting northing / dist=normal link=id
        ddfm=satterth;
    random rep;
    random rep*plot;
    random rep*plot*apass;
    random rep*plot*apass*hpass;
    random easting northing / type=rsmooth  knotinfo  knotmethod=kdtree(bucket=100);
run;
ods output close;

proc gplot data=knotinfo;
    plot northing*easting;
run;
quit;
```

FIG. 6-23. Knots automatically selected by GLIMMIX when the unrotated *easting* and *northing* coordinates are used in Example 6.2. There are knots included that are outside the convex hull of the field (i.e., outside of the field boundaries).

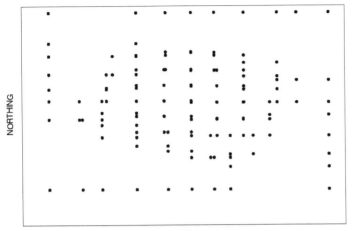

the tree correspond to the locations of the knots. Because they are used to partition the dataset into subsets of roughly equal size, the knots can be thought of as a multivariate generalization of univariate sample quantiles. When the option *knotmethod = kdtree* is used, the number of knots used is indirectly controlled using the *bucket* option. The *bucket = 100* option in Fig. 6–22 instructs GLIMMIX to select knots so that the dataset is partitioned into subsets, each with approximately 100 observations.

Figure 6–23 shows the locations of the radial smoother knots produced by GLIMMIX. For this dataset, specifying a bucket size of 100 results in GLIMMIX using 130 knots for the radial smoother. Notice that, in addition to placing knots within the boundary of the field, GLIMMIX has also placed several knots well outside its boundary. Apparently when using the kd-tree method of knot selection, the default behavior of GLIMMIX is not constrained to placing knots within the boundary around the observed data.

The problem is that the edges of the field are not parallel with the *northing* and *easting* coordinate axes. Rotating the coordinate axes first so that they align with the edges of the field would be one way to resolve this issue. The default behavior of the kd-tree method of knot selection then might give a reasonable set of knots. Alternatively, once we rotate the coordinate axes we could use the *knotmethod = equal* option to place knots uniformly across a grid spanning the observed data. This second option makes sense for this dataset since the observations are very uniformly distributed across the field, the rows being evenly spaced and observations within rows being taken every 2 seconds. Therefore, the axes were rotated so that they were aligned with the edges of the field. In addition, the rotated axes are shifted so that the southwest corner of the field is at the origin. Figure 6–24 shows

FIG. 6-24. Plot of the observed yield points after rotation and shifting of the axes to align them with the edges of the field in Example 6.2. The absence of points in the upper left corner of the rectangle corresponds to the location of a building.

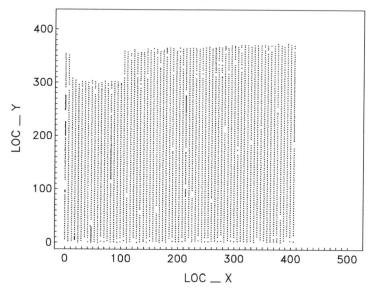

the locations of the yield points when plotted using the new set of axes. The new location variables are named *loc_x* and *loc_y*.

Figure 6–25 shows the GLIMMIX statements that fit a model using a radial smoother with knots that are uniformly spaced across a rectangular lattice that covers the observed data locations as given by the newly created location coordinates (*loc_x, loc_y*). When this option is used, the number of knots produced is the product of the macro variables &*nx* and &*ny*, where &*nx* is the number of lattice points in the *x* direction and &*ny* is the number of lattice points in the *y* direction.

How many knots should be used? That is, what values of &*nx* and &*ny* should be used? There is no easy answer to this question. The general consensus seems to be that using fewer knots than are needed can lead to inferential problems, while using more knots than are needed does not. However, increasing the number of knots increases the computational resources required to fit the model. With this in mind, one would be inclined to use the model with the smallest number of knots considered to be adequate. The question remains as to how many knots should be used to adequately account for the spatial variability in this dataset and allow us to make valid inferences about the fixed effects. Ruppert et al. (2003) recommended using at least 20 but no more than 150 knots when using a radial smoothing spline.

Our approach will be to fit the above model with different numbers of knots and then choose the number of knots to be used based on the following criteria: model fit, stability of the inferences regarding the fixed effects, and whether or not

FIG. 6-25. GLIMMIX statements to fit the model with radial smoother and knots selected on a uniform rectangular grid after rotation and shifting of the axes in Example 6.2.

```
/* First run of GLIMMIX with the nofit option to get knots on a uniform grid of size &nx
   and &ny. */

proc glimmix data=yield nofit ;
   class ec_zone nrate nem rep plot apass hpass ;
   model yld = ec_zone | nrate | nem loc_x loc_y;
   random rep;
   random rep*plot;
   random rep*plot*apass;
   random rep*plot*apass*hpass;
   random loc_x loc_y / type=rsmooth knotmethod=equal(&nx,&ny) knotinfo ;
run;


/* A data step to remove the knots in the building location would be placed here. */

/* GLIMMIX analysis using the knots from the knotinfo data file created above. */

proc glimmix data=yield plots=(studentpanel(conditional));
   class ec_zone nrate nem rep plot apass hpass ;
   model yld = ec_zone | nrate | nem loc_x loc_y / dist=normal link=id ddfm=satterth;
   random rep;
   random rep*plot;
   random rep*plot*apass;
   random rep*plot*apass*hpass;
   random loc_x loc_y / type=rsmooth knotmethod=data(knotinfo) knotinfo ;
   nloptions gconv=0;
run;
```

the model adequately accounts for spatial variation. The statistical model above was fit using the following 14 numbers of knots: 35, 62, 96, 138, 187, 216, 277, 308, 384, 504, 600, 699, 858, and 982. These numbers were obtained by placing knots uniformly across a square grid superimposed over the field and then removing those knots from the upper left portion of the field where no yield data were available (the location of a structure).

For the first criterion above for selecting the number of knots, which measure of model fit should be used? The Akaike information criterion (AIC) is a popular measure for comparing models that takes into account the number of parameters in the model. However, knots are random variables, not parameters, and the AIC does not take into account the number of knots in the radial smoother. Hence, the AIC is not useful here. One measure of fit that does take into account both the number of knots and the number of parameters in the model is the generalized cross-validation criterion (GCV) (Ruppert et al., 2003). The GCV criterion is an approximation to the usual cross-validation criterion and is interpreted in the same way. In particular, the GCV can be considered a function of the number of knots. The idea is to pick the number of knots that minimizes the GCV, if such a number exists. If such a number does not exist, then we can choose a number beyond which the per-knot reduction in GCV is small. The GCV is given by Ruppert et al. (2003) as

$$GCV(k) = \frac{RSS(k)}{\left[1 - n^{-1}Tr(S_k)\right]^2} = \frac{RSS(k)}{\left[Radial\_smoother\_df(Res)/n\right]^2},$$

where n is the number of observations, $RSS(k)$ is the residual sum of squares when k knots are used, S_k is the smoother matrix when k knots are used, and $Tr(S_k)$ is the trace of the smoother matrix. The second expression for GCV above is given because the value of $Radial\_smoother\_df(Res)$ in its denominator is given in the GLIMMIX output. This value was saved in a SAS dataset using an *ods* statement and used subsequently to calculate GCV.

Figure 6–26 presents a plot of GCV versus the number of knots used in the model. When more than 982 knots were used, GLIMMIX ran out of memory. Thus, 982 was the largest number of knots considered. It is clear that the GCV decreases as the number of knots increases, but over the range of numbers of knots considered GCV did not begin to increase. However, the marginal decrease in GCV is attenuated at larger numbers of knots. While the GCV is smallest when 982 knots are used, the per-knot decrease in GCV is relatively small when more than 384

FIG. 6–26. Plot of the generalized cross validation (GCV) statistic versus the number of knots used in the radial smoother in Example 6.2.

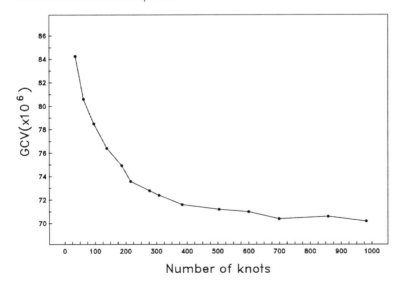

Number of knots

FIG. 6-27. Plot of the *p*-values for the tests of the fixed effects versus the number of knots used in the radial smoother in Example 6.2.

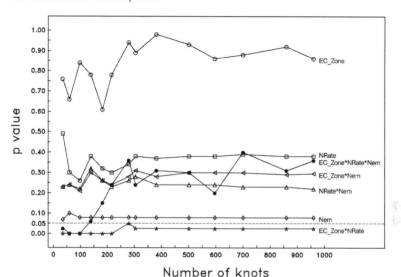

Number of knots

knots are used. Hence, based on this plot, it would appear that using 384 or more knots would be adequate from the standpoint of model fit.

The second criterion involves the stability of the inferences regarding the fixed effects. As the number of knots increases, the models account for varying degrees of spatial variability. As a result, the inferences regarding the fixed effects can change. Figure 6–27 shows a plot of the *p*-values for the various fixed effects in the model versus the number of knots used in the radial smoother. As the number of knots increases, several of the *p*-values are somewhat unstable until the number of knots reaches around 384. Beyond this, the *p*-values tend to settle down. Qualitatively, none of the inferences change once the number of knots reaches 384. Therefore, based on the stability of the inferences for the fixed effects, 384 knots appears to be adequate.

The last criterion involves adequately accounting for the spatial variation present in the data. Empirical semivariograms of the conditional studentized residuals were constructed for each of the 14 models. Based on inspection of the semivariogram plots, the radial smoothers in each of the models using 216 knots or more appear to account for most, if not all, of the residual spatial variation. For the purpose of adequately accounting for spatial variation, use of 216 knots appears to be adequate. Figure 6–28 shows the estimated semivariogram of the conditional studentized residuals when 216 knots are used.

Taking these results together, we conclude that 384 is the minimum number of knots that would adequately address all three criteria. The model using a radial smoother with 384 knots will be used in the remainder of this example. Figure 6–29 shows the knot placement relative to the rotated axes for the model using 384 knots.

FIG. 6-28. Plot of the empirical semivariogram of the conditional residuals for the model using a radial smoother with 216 knots in Example 6.2.

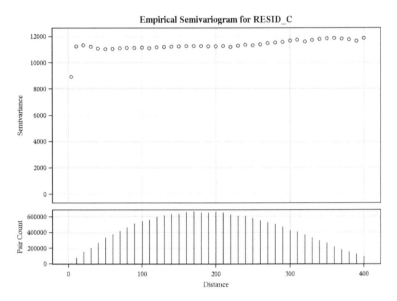

FIG. 6-29. Knot locations for a radial smoother using 384 knots in Example 6.2.

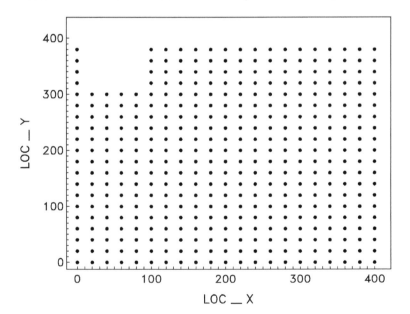

FIG. 6-30. GLIMMIX output containing the dimension information for the fitted model using a radial smoother with 384 knots in Example 6.2.

Dimensions	
G-side Cov. Parameters	5
R-side Cov. Parameters	1
Columns in X	50
Columns in Z	512
Subjects (Blocks in V)	1
Max Obs per Subject	6008

FIG. 6-31. GLIMMIX output containing the estimated covariance parameters for the model using a radial smoother with 384 knots in Example 6.2.

Covariance Parameter Estimates		
Cov Parm	Estimate	Standard Error
rep	1128.52	2413.59
rep*plot	233.54	485.90
rep*plot*apass	7.08E-17	.
rep*plot*apass*hpass	1214.17	276.71
var[rsmooth(loc_x, loc_y)]	5.5885	0.7262
residual	11330	216.17

The GLIMMIX output in Fig. 6–30 gives the dimensions of the X and Z design matrices and the number of covariance parameters in the model. In the original model without the radial smoother, there were 128 columns in the Z matrix, corresponding to the 128 random effects in the model (Fig. 6–16). Using a radial smoother with 384 knots adds 384 random effects to the original model, for a total of 512 random effects. Hence, there are 512 columns in the Z matrix. The model without the radial smoother has four G-side covariance parameters. The model with the radial smoother has one additional G-side covariance parameter, the variance of the radial smoother, for a total of five G-side covariance parameters.

The covariance parameter estimates for this model are given in Fig. 6–31. Note that for this model the estimate of the variance of the *rep* random effect is non-zero, although it is not significantly greater than zero. In addition, there does not appear to be significant variation among plots within replications. The estimate of the variance of the random effect for application passes within plots, while positive, is extremely close to zero. In this case the accompanying standard error of this estimate is missing because of computational underflow or overflow in the calculation of the estimate. On the other hand, there still appears to be significant variation between the harvest passes within application passes. There also appears to be sig-

FIG. 6-32. GLIMMIX output displaying plots of the conditional studentized residuals for the model using a radial smoother with 384 knots in Example 6.2.

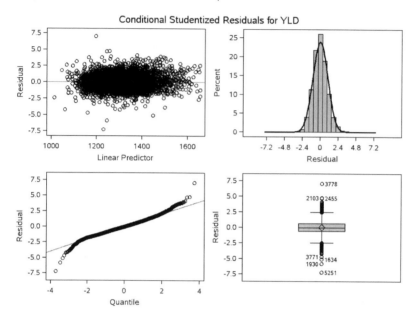

nificant variability associated with the radial smoother. Finally, by including the radial smoother in the model the residual variance has been reduced by nearly 28%

Figure 6–32 contains the panel of conditional studentized residual plots. As was the case for the first model considered, the empirical distribution of the residuals has somewhat heavier tails than that of the normal distribution, but otherwise the normality assumption does not appear to be violated to a great extent. Figure 6–33 shows an empirical semivariogram for the conditional studentized residuals from this model. The semivariogram is fairly flat from lag class one onward and across these lag classes is close to the estimated error variance of 11,330 obtained by the model. Within lag class zero, the value of the semivariogram is slightly less, being approximately 8900. This corresponds to an estimated spatial correlation of 0.217 (R^2 = 0.047), a negligible level of correlation. The estimated spatial correlation between residuals in lag classes one through four is no more than approximately 0.070 (R^2 = 0.0049), clearly a negligible amount. By incorporating the radial smoother in the model, the spatial correlation between residuals in lag class zero has been reduced by roughly 48% of that from the initial model (73% reduction in R^2), and the correlation between residuals farther apart has been rendered negligible. Hence adjacent residuals in the same harvest pass have a negligible or at most a weak level of correlation between them. Non-adjacent residuals in the same harvest pass and residuals in different harvest passes are essentially uncorrelated. These conditions are much more in line with the assumption of independent errors than was the case for the initial model. Therefore we

FIG. 6-33. Plot of the empirical semivariogram of the conditional residuals for the model using a radial smoother with 384 knots in Example 6.2.

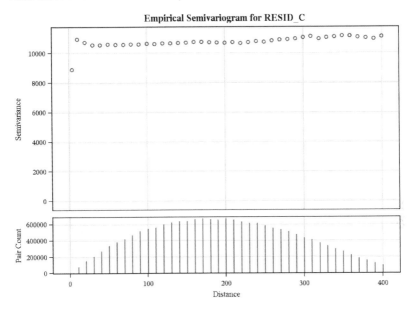

FIG. 6-34. GLIMMIX output containing tests of the fixed effects parameters for the model using a radial smoother with 384 knots in Example 6.2.

Type III Tests of Fixed Effects

Effect	Num DF	Den DF	F Value	Pr > F
ec_zone	2	4077	0.02	0.9767
nrate	2	6.03	1.13	0.3844
ec_zone*nrate	4	3155	1.15	0.3297
nem	1	5.372	4.42	0.0856
ec_zone*nem	2	3140	3.97	0.0189
nrate*nem	2	8.882	1.43	0.2897
ec_zone*nrate*nem	4	4010	1.14	0.3349
loc_x	1	119.3	0.01	0.9183
loc_y	1	118.9	0.05	0.8229

have much more confidence in the inferences we can draw about the fixed effects using this model.

Figure 6–34 summarizes the tests of the fixed effects in the model. The three-way interaction between *ec_zone*, *nrate*, and *nem* is not significant. The two-way interaction between *ec_zone* and *nrate*, which was significant in the model without the radial smoother, is not significant in this model. In fact, none of the effects involving *nrate* are significant. Hence, *nrate* does not appear to be having much of

FIG. 6-35. GLIMMIX statements to fit the initial model in Example 6.2 using alternative coding for the *random* statements that allows the covariance between harvest passes within application passes to be negative.

```
proc glimmix data=yield plots=(studentpanel(conditional));
    class ec_zone nrate nem rep plot apass hpass ;
    model yld = ec_zone | nrate | nem loc_x loc_y / dist=normal link=id
    ddfm=satterth;
    random intercept / subject=rep;
    random intercept / subject=rep*plot;
    random hpass / type=cs subject=rep*plot*apass;
    nloptions gconv=0;
run;
```

FIG. 6-36. GLIMMIX output containing estimates of the covariance parameters for the initial model in Example 6.2 based on alternative coding for the random effects.

Covariance Parameter Estimates			
Cov Parm	Subject	Estimate	Standard Error
Intercept	rep	0	.
Intercept	rep*plot	842.32	557.72
Variance	rep*plot*apass	1161.69	319.22
CS	rep*plot*apass	96.3381	305.37
Residual		15649	287.53

an impact on cotton lint yield on this particular field. On the other hand, the two-way interaction between *ec_zone* and *nem* remains significant in this model. The results of these tests indicate that the effectiveness of the nematicide depends on the EC_a zone in which it is being applied, and therefore in developing a prescription for the nematicide treatment its levels should be compared within each EC_a zone. The prescription for the nitrogen treatment can be a blanket treatment.

Recall that in the analysis using the radial smoother the estimate of the variance of the random effect of application passes with plots is extremely close to zero. Note that this variance component is, more generally, the covariance between the two harvest passes within a given application pass, and covariances can be negative. Modeling this parameter as a variance component using the random statement in Fig. 6–25 imposes a positivity constraint that prevents it from being negative. It is possible that the covariance estimate is bumping up against the zero boundary on the positive side because the actual covariance is negative, but the positivity constraint imposed by the coding will not let the estimate take a negative value. It is possible to recode this random statement so that this covariance parameter can take negative values. This has been done in Fig. 6–35 for the model without the radial smoother and in Fig. 6–38 for the model with

FIG. 6-37. GLIMMIX output containing the tests of the fixed effects for the initial model in Example 6.2 based on alternative coding for the random effects.

Type III Tests of Fixed Effects

Effect	Num DF	Den DF	F Value	Pr > F
ec_zone	2	5947	165.21	<0.0001
nrate	2	10.59	0.47	0.6389
ec_zone*nrate	4	5925	4.62	0.0010
nem	1	10.54	3.27	0.0990
ec_zone* nem	2	5913	9.19	0.0001
nrate* nem	2	10.56	0.06	0.9409
ec_zone*nrate* nem	4	5927	2.14	0.0727
loc_x	1	10.96	9.57	0.0103
loc_y	1	5952	23.31	<0.0001

FIG. 6-38. GLIMMIX statements to fit the model with radial smoother and knots selected on a uniform rectangular grid after rotation and shifting of the axes in Example 6.2 using alternative coding for the *random* statements that allows the covariance between harvest passes within application passes to be negative.

```
/* First run of GLIMMIX with the nofit option to get knots on a uniform grid of size &nx and
   &ny. */

proc glimmix data=yield nofit ;
   class ec_zone nrate nem rep plot apass hpass ;
   model yld = ec_zone | nrate | nem loc_x loc_y;
   random intercept / subject=rep;
   random intercept / subject=rep*plot;
   random hpass / type=cs subject=rep*plot*apass;
   random loc_x loc_y / type=rsmooth  knotmethod=equal(&nx,&ny) knotinfo ;
run;

/* A data step to remove the knots in the building location would be placed here. */

/* GLIMMIX analysis using the knots from the knotinfo data file created above. */

proc glimmix data=yield plots=(studentpanel(conditional));
   class ec_zone nrate nem rep plot apass hpass ;
   model yld = ec_zone | nrate | nem loc_x loc_y / dist=normal  link=id ddfm=satterth;
   random intercept / subject=rep;
   random intercept / subject=rep*plot;
   random hpass / type=cs subject=rep*plot*apass;
   random loc_x loc_y / type=rsmooth  knotmethod=data(knotinfo) knotinfo ;
   nloptions gconv=0;
run;
```

FIG. 6-39. GLIMMMIX output containing the estimated covariance parameters for the model using a radial smoother with 384 knots in Example 6.2 based on alternative coding for the random effects.

Covariance Parameter Estimates

Cov Parm	Subject	Estimate	Standard Error
intercept	rep	1226.52	2579.27
intercept	rep*plot	280.42	559.41
variance	rep*plot*apass	1245.53	331.32
cs	rep*plot*apass	−61.2698	315.69
var[rsmooth(loc_x, loc_y)]		5.5859	0.7258
residual		11330	216.18

FIG. 6-40. GLIMMIX output containing tests of the fixed effects parameters for the model using a radial smoother with 384 knots in Example 6.2 based on alternative coding for the random effects.

Type III Tests of Fixed Effects

Effect	Num DF	Den DF	F Value	Pr > F
ec_zone	2	4074	0.02	0.9769
nrate	2	5.706	1.15	0.3795
ec_zone*nrate	4	3154	1.15	0.3302
nem	1	5.034	4.18	0.0959
ec_zone* nem	2	3139	3.98	0.0187
nrate* nem	2	8.44	1.48	0.2805
ec_zone*nrate* nem	4	4008	1.15	0.3312
loc_x	1	119.3	0.01	0.9208
loc_y	1	118.9	0.05	0.8227

the radial smoother. Figure 6–36 shows the covariance parameter estimates for the model without the radial smoother. These estimates, along with their standard errors, are essentially identical to the covariance parameter estimates in Fig. 6–17 from the original model without the radial smoother. In particular, for this analysis the covariance parameter estimate for application pass does not take a negative value. Figure 6–37 shows the tests of fixed effects for the model without the radial smoother using alternative random effect coding. The p-values are identical to those in Fig. 6–18 from the original model without the radial smoother, as expected since the covariance parameters have not changed. Figure 6–39 shows the covariance parameter estimates for the model with the radial smoother using the alternative random effect coding. Note that in this case the covariance parameter estimate for application pass does in fact take a negative value, although it does not appear to be significantly different than zero. The rest of the covariance parameter estimates are very similar to those in Fig. 6–31. Figure 6–40 gives the tests of fixed effects for the model with the radial smoother using the alternative random effect coding. The p-values are very close to the corresponding values in Fig. 6–34 obtained from the model with the radial smoother and the original random statement coding. For both sets of models, therefore, the inferences are the same.

This example illustrates the use of radial smoothers in GLIMMIX to account for spatial variability and thereby reduce and/or eliminate residual spatial correlation that results from unaccounted for spatial trend. This has ramifications for the inferences being made about the field treatments under consideration and the amount of work that would subsequently be required to develop a treatment prescription based on the results of the model. Without the radial smoother the residuals exhibited significant spatial correlation, calling into question the assumption of independent errors imposed by the initial model. The dataset was too large to fit a more general model with a parametric spatial covariance structure that would allow for errors to be correlated, a common problem with datasets generated by precision agriculture applications. Using a radial smoother to account for residual spatial trend reduced spatial correlation to negligible levels and resulted in the assumption of independent errors being plausible. This had an impact on the inferences about the treatments being considered. Results from the model without the radial smoother indicated that a variable rate treatment prescription would be needed for both nitrogen and the nematicide treatment, whereas results from the model with the radial smoother indicated that a variable rate treatment prescription would be needed for only the nematicide. ■

REFERENCES CITED

Burris, E., D. Burns, K.S. McCarter, C. Overstreet, M.C. Wolcott, and E. Clawson. 2010. Evaluation of the effects of Telone II (fumigation) on nitrogen management and yield in Louisiana delta cotton. Precis. Agric. 11:239–257. doi:10.1007/s11119-009-9129-x

Cressie, N.A.C. 1993. Statistics for spatial data. Revised ed. John Wiley and Sons, New York.

Galecki, A.T. 1994. General class of covariance structures for two or more repeated factors in longitudinal data analysis. Comm. Statist. Theory Methods 23:3105–3119. doi:10.1080/03610929408831436

Lennsen, A.W., G.D. Johnson, and G.R. Carlson. 2007a. Cropping sequence and tillage system influences annual crop production and water use in semiarid Montana, USA. Field Crops Res. 100:32–43. doi:10.1016/j.fcr.2006.05.004

Lennsen, A.W., J.T. Waddell, G.D. Johnson, and G.R. Carlson. 2007b. Diversified cropping systems in semiarid Montana: Nitrogen use during drought. Soil Tillage Res. 94:362–375. doi:10.1016/j.still.2006.08.012

Littell, R.C., G.A. Milliken, W.W. Stroup, R.D. Wolfinger, and O. Schabenberger. 2006. SAS for mixed models. 2nd ed. SAS Institute, Cary, NC.

McCarter, K., and E. Burris. 2010. Accounting for spatial correlation using radial smoothers in statistical models used for developing variable-rate treatment prescriptions. *In* Proceedings of the 10th International Conference on Precision Agriculture. Denver, CO.

Roy, A., and R. Khattree. 2005. Discrimination and classification with repeated measures data under different covariance structures. Commun. Stat. Simul. Comput. 34:167–178. doi:10.1081/SAC-200047072

Ruppert, D., M.P. Wand, and R.J. Carroll. 2003. Semiparametric regression. Cambridge Univ. Press, New York.

Schabenberger, O., and C.A. Gotway. 2005. Statistical methods for spatial data analysis. Chapman and Hall/CRC, Boca Raton, FL.

CHAPTER 7

DESIGNING EXPERIMENTS

7.1 INTRODUCTION

In this chapter the use of generalized linear mixed models as a planning tool for the design of agronomic experiments is discussed. The reader might well ask, "Don't generalized linear mixed models concern modeling and data analysis? What does this have to do with design?" To quote from the classic text *Experimental Designs* (Cochran and Cox, 1992), "It has come to be recognized that the time to think about statistical inference, or to seek [a statistician's] advice, is when the experiment is being planned." Hahn (1984) put it more forcefully, "Statisticians make their most valuable contributions if they are consulted in the planning stages of an investigation. Proper experimental design is often more important than sophisticated statistical analysis." He continues, quoting H. Ginsburg as saying, "When I'm called in after it's all over, I often feel like a coroner. I can sign the death certificate—but do little more." Light et al. (1990) stated it slightly differently, "You cannot save by analysis what you bungle by design."

In his text *The Design of Experiments*, Mead (1988) noted that the development of experimental design concepts was "restricted by the earlier need to develop mathematical theory for design in such a way that the results from the design can be analyzed without recourse to computers." Because of the increasing sophistication of statistical modeling and the dramatic increase in capacity of modern computers, Mead argued, "The fundamental concepts now require reexamination and re-interpretation outside the limits of classical mathematical theory so that the full range of design possibilities may be considered."

Following his line of thought, while generalized linear mixed models provide researchers with expanded flexibility to apply regression and analysis of variance approaches to data that are not normally distributed, conventional wisdom about the design of experiments reflects the "restraints" referred to by Mead. For researchers to genuinely benefit from generalized linear mixed models, experiments must be designed to allow their full potential to be realized. This is done by using generalized linear mixed model power, precision, and sample-size analysis in the planning process.

doi:10.2134/2012.generalized-linear-mixed-models.c7

Analysis of Generalized Linear Mixed Models in the Agricultural and Natural Resources Sciences
Edward E. Gbur, Walter W. Stroup, Kevin S. McCarter, Susan Durham, Linda J. Young, Mary Christman, Mark West, and Matthew Kramer

As an example of an area where this type of pre-experiment preparation is rigorously followed, consider the pharmaceutical industry. Regulations require that investigators finalize study protocols before their commencement. A protocol must describe the design of the study, identify and rank, in order of importance, the various hypotheses to be tested, and specify the models to be fit and the statistical methods to be used in performing the analyses. As part of these preparations, power analyses are conducted to ensure that the study will be adequate for its intended purpose. This is very important. Even aside from financial considerations, it would be unethical to expose subjects to the potential risks of a clinical trial without ensuring a reasonable chance of detecting a clinically relevant treatment effect. In addition, it is undesirable to expose more subjects to the potential risks than are necessary to obtain a specified level of power.

This level of pre-experiment preparation is not, and may never be, required of researchers in most academic fields. However, it can be considered a "best practice model," a goal to strive for. In fact, we are seeing a movement in this direction in several fields. For example, grant-funding agencies such as NIH now require that power analyses be included in grant proposals. Even when not formally required, including a power analysis gives a grant proposal a competitive advantage because it shows funding agencies that the researcher has thought carefully about the proposed design and its potential to obtain results. In all cases, it is in the researcher's enlightened self-interest to assess the power and precision of a proposed design before data collection begins. This is especially true when generalized linear mixed models are to be used to analyze the data. A design that is optimal for analysis of variance or regression with normally distributed data may be unsuitable for non-normal data such as counts, percentages, and times to an event. What reasonable researcher would invest time, effort, and money in an experiment without first getting an idea of the likelihood of successfully detecting scientifically relevant results, should they exist?

The purpose of this chapter is to show how generalized linear mixed model based tools can be used in planning experiments that will be analyzed using generalized linear mixed models. Specifically, we show how generalized linear mixed models can be used to assess the expected power profile and the precision of a proposed experiment of a given size and type, and to guide modifications when they are necessary. In many cases, a given set of treatments and a given number of experimental units can be arranged into more than one plausible design, often with very different power profiles with respect to the researcher's objectives. Power and precision analysis can be used to assess the strengths and drawbacks of competing designs. The tools presented in this chapter should be considered essential in planning agronomic experiments and experiments in other fields as well.

7.2 POWER AND PRECISION

Power is defined as the probability of rejecting the null hypothesis when in fact the null hypothesis is false and therefore should be rejected. In practical terms, the null hypothesis states that a given treatment has no effect, while the research or alternative hypothesis states that a treatment does indeed have an effect. Hence, power is the probability that one will be able to demonstrate the credibility of the research hypothesis, with acceptable scientific rigor, when the research hypothesis is in fact true.

Power analysis is, in essence, the computation of that probability. Specifically, one determines the minimum treatment effect one considers to be scientifically relevant and then computes the probability that a proposed design will show that difference to be statistically significant. Precision analysis is similar, but instead of focusing on power, one determines how wide a confidence interval for the treatment effect is expected to be for the proposed design.

7.3 POWER AND PRECISION ANALYSES FOR GENERALIZED LINEAR MIXED MODELS

The first step in conducting a power and a precision analysis is to identify the nature of the response variable, its distribution, and the effect(s) of interest. For example, in a one factor, completely randomized design, the model describing the treatment effect is $\beta_0 + T_i$, $i = 1, \ldots, t$, where T_i is the effect of the ith treatment and β_0 is the intercept or overall mean. For normally distributed response variables, $\beta_0 + T_i$ directly models the treatment mean μ_i. For binomial responses, $\beta_0 + T_i$ usually models the logit of π_i, where π_i denotes the probability of the occurrence of the event of interest (success) for the ith treatment. For counts modeled by an appropriate counting distribution, $\beta_0 + T_i$ models $\log(\mu_i)$, where μ_i is the expected count for the ith treatment.

The hypotheses to be tested are specified in terms of treatment differences or, more generally, contrasts $\Sigma_i k_i T_i$, where the k_i are constants chosen to define the effect of interest. Under the null hypothesis, H_0: $\Sigma_i k_i T_i = 0$ and under the research hypothesis, H_A: $\Sigma_i k_i T_i \neq 0$. For example, setting $k_1 = 1$, $k_2 = -1$, and the remaining $k_i = 0$ defines the contrast $T_1 - T_2$, the difference between treatments 1 and 2. In this case, H_0: $T_1 - T_2 = 0$ and H_A: $T_1 - T_2 \neq 0$. A generalized linear mixed model test of this hypothesis is based on an F statistic. If H_0 is true, this statistic has an approximate central F distribution, denoted $F_{(0, \text{Ndf}, \text{Ddf})}$, where Ndf denotes the numerator degrees of freedom, and Ddf denotes the denominator degrees of freedom. Under the research hypothesis, the F statistic has an approximate non-central F distribution, denoted by $F_{(\varphi, \text{Ndf}, \text{Ddf})}$, where φ denotes the non-centrality parameter. Without going into technical details, the non-centrality parameter depends on the quantity

$$\left(\text{sample size} \right) \times \left(\sum_i k_i T_i \right) \Big/ \left(\text{variance of treatment effect} \right)$$

A formal definition and technical details can be found in experimental design textbooks, for example, Hinkelmann and Kempthorne (1994). Note that under the null hypothesis, $\Sigma_i k_i T_i = 0$, and hence, the non-centrality parameter φ is also 0. Under the research hypothesis, $\Sigma_i k_i T_i > 0$, and hence, $\varphi > 0$. The non-centrality parameter

FIG. 7-1. The effect of the non-centrality parameter on the F distribution.

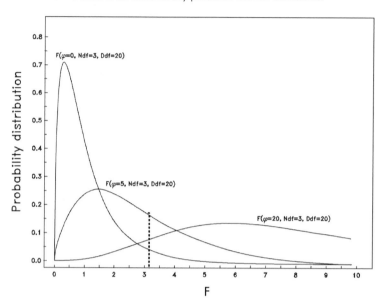

increases when either the effective sample size increases, the treatment effect increases, or the variance of the treatment effect decreases.

Figure 7–1 illustrates the effect of the non-centrality parameter on the F distribution. In the figure, the central F is the highly right-skewed distribution in the left-most position and represents the distribution of the test statistic under the null hypothesis. The dashed vertical line represents the critical value of the test for $\alpha = 0.05$. An observed value of the F statistic greater than the critical value would lead to rejection of the null hypothesis. The two non-central F distributions show what happens as the non-centrality parameter increases; namely, the larger the value, the more the distribution is shifted to the right. The area under the curve to the right of the critical value corresponds to the power of the test. As φ increases, the power of the test increases.

Precision analysis is based on interval estimation of the effect of interest, $\Sigma_i k_i T_i$. The ratio of the estimated contrast to its standard error has an approximate t distribution with Ddf degrees of freedom. Thus, a $100(1 - \alpha)\%$ confidence interval for $\Sigma_i k_i T_i$ is of the form

$$\text{estimate of } \sum_i k_i T_i \pm t_{(\alpha,\,\text{Ddf})} \times \text{standard error of } \sum_i k_i T_i$$

For a given design, one can use generalized linear mixed model software to compute the approximate standard error and hence, the expected confidence interval width for the contrast.

7.4 METHODS OF DETERMINING POWER AND PRECISION

There are two primary ways of evaluating the power and precision of an experiment using generalized linear mixed model software. The first method, henceforth referred to as the probability distribution method, is applicable when we know (or can approximate) the sampling distribution of the test statistic under the conditions of the research hypothesis. In this case, one determines the non-centrality parameter of the distribution of the test statistic at a particular point under the research hypothesis. One then approximates the power of the test using the area under this non-central distribution to the right of the critical value, as illustrated in Fig. 7–1. One can use GLIMMIX in conjunction with SAS's (SAS Institute, Cary, NC) probability functions to perform these calculations.

The second method uses simulation to estimate the power of a test and is applicable regardless of whether we know or can approximate the actual sampling distribution of the test statistic under the research hypothesis. All that is necessary to use this method is the ability to perform the test of interest and generate random numbers from the distribution of interest. To estimate power via simulation, one uses a random number generator to create a large number of independent data sets that match the proposed study design and reflect the conditions under the research hypothesis to be detected. The SAS data step and random number functions can be used to create these data sets. One then performs the desired generalized linear mixed model analysis for each data set separately, in each case keeping track of whether the null hypothesis has been rejected. Since the simulated samples are independent, the number of samples for which the null hypothesis is rejected has a binomial distribution with the number of trials equal to the number of simulated datasets and with probability equal to the true power of the hypothesis test. This fact provides a basis for making inferences about the true power of the test, including computing point and confidence interval estimates and testing hypotheses about the power. In particular, the proportion of simulated datasets for which the null hypothesis is rejected gives a point estimate of the power of the test. In addition, a confidence interval for the power of the test can be computed from these simulation results. For precision analysis, one calculates the mean and variance of the width of the confidence intervals over all simulated samples produced by the generalized linear mixed model. GLIMMIX can be used for both power analysis and precision analysis using the simulation method.

A major advantage of the probability distribution method over the simulation method is that it is quicker and easier to set up, allowing rapid comparison of competing designs, different effect sizes, different levels of variation, or different sample sizes. However, to use this method one must know (or know an approximation of) the actual sampling distribution of the test statistic. One advantage of the simulation method over the probability distribution method is that it is applicable for any design and any type of analysis, regardless of whether the behavior of the test statistic is well understood. The only requirement is that one is able to generate data according to the study design.

A second advantage of the simulation method is that, since it involves analyzing hundreds (or thousands) of datasets similar to those that are expected from the study, it allows one to see exactly what the analysis will look like and how the GLIMMIX procedure will behave with data from the proposed design. The simulation method may reveal any troublesome behavior GLIMMIX may display for a contemplated design. Researchers can use such fair warning to make needed changes in the proposed design before the data are collected and it is too late. One disadvantage of the simulation method is that, because it requires analyzing a large number of samples, it can be much more time consuming, especially when evaluating power over a wide range of possibilities under the alternative hypothesis. A benefit of both methods is that the programs used to perform the power analysis can be used later, perhaps with minor alterations, to analyze the real data once they have been obtained.

The approach taken in this chapter and recommended for use in practice is to use the probability distribution method to compare the various design alternatives for a study and to identify one or more that provide the desired power characteristics. Then use the simulation method to verify the power approximations obtained from the probability distribution method. Again we emphasize that all of this should be done during the planning stages of an experiment, before data collection starts.

Four items of information are required to perform a power analysis:

- the minimum treatment effect size $\sum_i k_i T_i$ considered scientifically relevant,

- the assumed probability distribution of the response variable,

- an approximate idea of the magnitude and nature of the variation and correlation present in the data,

- a clear idea of the structure of the proposed design.

A few clarifications about these required items are in order. First, providing the scientifically relevant treatment effect size does not mean knowing in advance how big a difference there will be among treatment means. Many researchers short-circuit power analysis by saying, "I can't give you that. If I knew how different the treatment means are, I wouldn't have to run the experiment!" True, but that is not the question. The question is, "Given your knowledge about the research question that is motivating this study, what is the minimum difference that would be considered important if, in fact, it exists?" Would a 1 kg ha^{-1} increase in yield be considered too trivial to matter? Would a 10 kg ha^{-1} difference be considered extremely important? What about a 5 kg ha^{-1} difference?

Second, to get an idea of the magnitude and nature of the variation and correlation present in the data, one must identify the relevant sources of variation (e.g., blocking, experimental unit error), distinguishing between whole plot and split plot variance if a split plot experiment is being proposed, characterizing likely correlation structures among measurements over time if a repeated measures design is proposed, and characterizing likely spatial variability if there is reason to believe it is present. Several of these issues will be addressed in the examples that follow.

If it appears from these requirements that a great deal of conversation between the researcher and statistical scientist should be occurring early in the planning of the experiment, then the reader has the right idea.

7.5 IMPLEMENTATION OF THE PROBABILITY DISTRIBUTION METHOD

This basic approach originated with linear models using PROC GLM in SAS (Littell, 1980; O'Brien and Lohr, 1984; Lohr and O'Brien, 1984). Stroup (1999) extended the method to linear mixed models using PROC MIXED. Stroup (2002) described the implementation of the probability distribution method for linear mixed models using PROC MIXED, focusing on experiments in the presence of spatial variation, and provided evidence of the accuracy of these methods via simulation. Littell et al. (2006) provided additional detail and examples for linear mixed models. In this section, the method is extended to generalized linear mixed models.

Implementation of the probability distribution method requires four basic steps. These steps are listed here and are illustrated by a simple example using a two-treatment, completely randomized design for a normally distributed response. The steps are as follows:

1. Create an "exemplary data set" (O'Brien and Lohr, 1984), that is, a data set whose structure is identical to the data that would be collected using the proposed design but with the observed data replaced by means reflecting the treatment difference to be detected under the research hypothesis.

2. Determine the numerator and denominator degrees of freedom and the non-centrality parameter that follows from the design and the research hypothesis. These can be obtained from the generalized linear mixed model software.

3. Determine the critical value based on the numerator and denominator degrees of freedom found in Step 2.

4. Compute the power, that is, the probability that the test statistic exceeds the critical value using the numerator and denominator degrees of freedom and the non-centrality parameter determined in Step 2 and the critical value found in Step 3.

EXAMPLE 7.1

Suppose we want to compare two treatments, a reference (or control) and an experimental treatment, using a completely randomized design in which the response is normally distributed. Suppose further that experience with the control treatment indicates it has a mean response of approximately 10 units with a standard deviation of roughly 10% of the mean. That is, for the control treatment $\mu = 10$ and $\sigma = 1$. The researcher believes that it would be scientifically relevant if the experimental treatment increases the mean response by 10% or more; i.e., to at least $\mu = 11$. The researcher wants to know the probability that four replications per treatment would show the scientifically relevant difference to be statistically significant. With this information, the probability distribution method is implemented as follows.

FIG. 7-2. SAS statements to create an
exemplary data set for Example 7.1.

FIG. 7-3. The exemplary data
set for Example 7.1 from the
PROC PRINT in Fig. 7-2.

```
data crd_example;
   input trt mu;
   do rep=1 to 4 by 1;
      output;
   end;
   datalines;
   0  10
   1  11
run;

proc print data=crd_example noobs;
   var trt mu rep;
run;
```

trt	mu	rep
0	10	1
0	10	2
0	10	3
0	10	4
1	11	1
1	11	2
1	11	3
1	11	4

Step 1. Create the exemplary data set.

This will have four lines of data per treatment (one per replication), each line containing the treatment and the mean for that treatment under the research hypothesis (10 for control, 11 for the experimental treatment).

The SAS data step to accomplish Step 1 is shown in Fig. 7–2, and the data file that it created is shown in Fig. 7–3. There are two input variables, *trt* (treatment) and *mu* (the mean for the treatment specified by *trt*). *trt* takes two values, 0 for the control and 1 for the experimental treatment, and *mu* takes the values 10 and 11, respectively, corresponding to the minimum scientifically relevant difference as specified by the researcher. The *do, output,* and *end* statements form a "do-loop" to create the required four lines of data per treatment.

Step 2. Analyze the exemplary dataset using GLIMMIX to obtain the terms needed to compute the power and the precision of the experiment.

The GLIMMIX statements for Step 2 are given in Fig. 7–4. The *class* and *model* statements are exactly as they would be when the actual data from the experiment are analyzed. The *parms* statement sets the error variance to 1. The *hold* option instructs the procedure to fix it at $\sigma^2 = 1$ and to not treat it as a parameter to be estimated. (The *parms* statement and *noprofile* option would be removed when analyzing the real data). The *diff* and *cl* options in the LSMEANS statement direct the procedure to compute the projected 95% confidence interval for the treatment difference. For this example, this is the precision analysis. The output shown in Fig. 7–5 gives the information needed for the precision analysis. The *ods* statement causes the GLIMMIX procedure to create a new data set, which we have named *power_terms*, that contains the various values needed for the power analysis (F value, numerator and denominator degrees of freedom). The contents of this file are shown in Fig. 7–6.

FIG. 7-4. GLIMMIX statements to compute terms needed for the power/precision analysis for Example 7.1.

```
proc glimmix data=crd_example noprofile;
    class trt;
    model mu=trt / dist=normal link=id;
    parms (1) / hold=1;
    lsmeans trt / diff cl;
    ods output tests3=power_terms;
run;

proc print data=power_terms noobs;
run;
```

FIG. 7-5. GLIMMIX output containing the information required for the precision analysis in Example 7.1.

trt Least Squares Means

trt	Estimate	Standard Error	DF	t Value	Pr > |t|	Alpha	Lower	Upper
0	10.0000	0.5000	6	20.00	<0.0001	0.05	8.7765	11.2235
1	11.0000	0.5000	6	22.00	<0.0001	0.05	9.7765	12.2235

Differences of trt Least Squares Means

trt	_trt	Estimate	Standard Error	DF	t Value	Pr > |t|	Alpha	Lower	Upper
0	1	−1.0000	0.7071	6	−1.41	0.2070	0.05	−2.7302	0.7302

FIG. 7-6. The contents of the *power_terms* file for Example 7.1 from the PROC PRINT in Fig. 7–4.

Effect	NumDF	DenDF	FValue	ProbF
trt	1	6	2.00	0.2070

From Fig. 7–5, the precision analysis shows that if this experiment is run with four replications, the expected standard error of a treatment mean will be 0.5, the expected standard error of a treatment difference will be 0.707, and the expected width of the 95% confidence interval for the treatment difference will be 0.730 − (−2.730) = 3.46 units.

Steps 3 and 4. The values in the data set created by the *ods* statement (*power_terms*) are used to obtain the critical value, compute the non-centrality parameter, and then evaluate the power.

The SAS statements to perform Steps 3 and 4 are shown in Fig. 7–7. These statements, perhaps with minor alterations, are used for all of the examples presented in this chapter. The data step creates a new data set called *power* from the data set *power_terms* produced by the GLIMMIX analysis. The non-centrality parameter under the research hypothesis is equal to the product of the numerator degrees of freedom (*NumDF*) and the F-value. In this example, α, the type I error probability, is set to 0.05. The critical value of F is calculated using the *finv* function. The statement shown obtains the critical value from the central F-distribution (i.e., F under the null hypothesis) using the numerator and denominator degrees of freedom provided by GLIMMIX as a result of analyzing the exemplary dataset. The *ProbF* function determines the area under the non-central F distribution (i.e., F under the research hypothesis) to the left of the critical value. Subtracting this area from one yields the power. The resulting information from the PROC PRINT statement appears in Fig. 7–8.

The approximated power of the proposed experiment is 0.2232. In other words, given the scientifically relevant difference specified above and the assumed magnitude of the error variance, the researcher has less than a one in four chance of obtaining data that will allow rejection of the null hypothesis. Clearly, four replications do not provide adequate power.

One can evaluate power for different numbers of replications by modifying the upper limit in the *do* statement in the creation of the exemplary data set. To find the minimum number of replications required to obtain a given power, one can

FIG. 7–7. SAS statements to compute power from the GLIMMIX output for Example 7.1.

```
data power;
    set power_terms;
    alpha = 0.05;
    NonCent_parm = NumDF*Fvalue;
    FCrit = Finv(1 - alpha, NumDF, DenDF, 0);
    Power = 1 - ProbF(FCrit, NumDF, DenDF, NonCent_parm);
run;


proc print data=power;
run;
```

FIG. 7–8. Power analysis results for Example 7.1 from the PROC PRINT in Fig. 7–5.

Obs	Effect	NumDF	DenDF	FValue	ProbF	alpha	NonCent_parm	FCrit	Power
1	trt	1	6	2.00	0.2070	0.05	2	5.98738	0.22319

progressively change this upper limit until the desired level of power is obtained. For example, suppose we wish to determine the smallest number of replications for which the test has power at least 0.80. Varying the upper endpoint in the *do* statement in this way, we find that 16 replications result in the power being 0.78, and 17 replications result in the power being 0.81. Therefore 17 is the minimum number of replications that will provide at least an 80% chance of detecting the treatment difference specified above, assuming an error variance of 1. The power provided by other numbers of replications when the error variance is 1 is given in Table 7–1 under the column labeled approximated power.

TABLE 7–1. Approximated and estimated power for the comparison of two treatments in a completely randomized design with variance equal to 1 in Example 7.1.

Number of replications	Approximated power†	Estimated power†	Lower confidence limit‡	Upper confidence limit‡
4	0.2232	0.2188	0.1934	0.2441
10	0.5620	0.5713	0.5410	0.6016
15	0.7529	0.7813	0.7559	0.8066
16	0.7814	0.7813	0.7559	0.8066
17	0.8070	0.7881	0.7631	0.8131
18	0.8300	0.8379	0.8153	0.8605
19	0.8506	0.8721	0.8516	0.8925
20	0.8690	0.8652	0.8443	0.8861
25	0.9337	0.9287	0.9130	0.9445
30	0.9677	0.9648	0.9536	0.9761
31	0.9721	0.9678	0.9570	0.9786
32	0.9760	0.9746	0.9650	0.9842
33	0.9793	0.9678	0.9570	0.9786
34	0.9822	0.9795	0.9708	0.9882
35	0.9848	0.9834	0.9756	0.9912
40	0.9930	0.9941	0.9895	0.9988
45	0.9968	0.9990	0.9971	1.0000
50	0.9986	0.9990	0.9971	1.0000
55	0.9994	0.9961	0.9923	0.9999
60	0.9997	0.9990	0.9971	1.0000
65	0.9999	1.0000	1.0000	1.0000
70	1.0000	1.0000	1.0000	1.0000

† Approximated power is based on the probability distribution method. Estimated power is based on the simulation method with 1024 simulated samples.

‡ 95% confidence limits for the estimated power.

What is the effect on power if we have underestimated the error variance? For example, how much power will 17 replications provide if the error variance σ^2 is actually 2, or even worse if it is as large as 4? This is easily answered by changing the *parms* statement in the GLIMMIX procedure. Re-running the procedure above with *parms (2)*, we see that with 17 replications the power drops to 0.52 when σ^2 = 2 (Table 7–2), and re-running with *parms (4)* shows that with 17 replications the power drops even further to 0.29 when σ^2 = 4 (Table 7–3). By increasing the number of replications as described above we see that if σ^2 were actually 4, it would take 65 replications to achieve power of 0.80 (Table 7–3).

TABLE 7–2 Approximated and estimated power for the comparison of two treatments in a completely randomized design with variance equal to 2 in Example 7.1.

Number of replications	Approximated power†	Estimated power†	Lower confidence limit‡	Upper confidence limit‡
4	0.1356	0.1357	0.1148	0.1567
10	0.3220	0.3408	0.3118	0.3699
15	0.4642	0.4551	0.4246	0.4856
16	0.4904	0.4932	0.4625	0.5238
17	0.5158	0.5029	0.4723	0.5336
18	0.5403	0.5547	0.5242	0.5851
19	0.5640	0.5684	0.5380	0.5987
20	0.5868	0.5908	0.5607	0.6209
25	0.6879	0.6885	0.6601	0.7168
30	0.7682	0.7578	0.7316	0.7841
31	0.7820	0.7715	0.7458	0.7972
32	0.7951	0.8008	0.7763	0.8252
33	0.8076	0.8008	0.7763	0.8252
34	0.8193	0.8135	0.7896	0.8373
35	0.8305	0.8262	0.8030	0.8494
40	0.8776	0.8711	0.8506	0.8916
45	0.9127	0.9199	0.9033	0.9365
50	0.9383	0.9424	0.9281	0.9567
55	0.9568	0.9531	0.9402	0.9661
60	0.9700	0.9678	0.9570	0.9786
65	0.9794	0.9795	0.9708	0.9882
70	0.9859	0.9873	0.9804	0.9942

† Approximated power is based on the probability distribution method. Estimated power is based on the simulation method with 1024 simulated samples.

‡ 95% confidence limits for the estimated power.

TABLE 7-3 Approximated and estimated power for the comparison of two treatments in a completely randomized design with variance equal to 4 in Example 7.1.

Number of replications	Approximated power†	Estimated power†	Lower confidence limit‡	Upper confidence limit‡
4	0.0923	0.0908	0.0732	0.1084
10	0.1851	0.1943	0.1701	0.2186
15	0.2624	0.2695	0.2424	0.2967
16	0.2777	0.2803	0.2528	0.3078
17	0.2930	0.2871	0.2594	0.3148
18	0.3081	0.3018	0.2736	0.3299
19	0.3231	0.3281	0.2994	0.3569
20	0.3379	0.3398	0.3108	0.3689
25	0.4101	0.4102	0.3800	0.4403
30	0.4779	0.4873	0.4567	0.5179
31	0.4909	0.4785	0.4479	0.5091
32	0.5036	0.4922	0.4616	0.5228
33	0.5162	0.5449	0.5144	0.5754
34	0.5285	0.5146	0.4840	0.5453
35	0.5407	0.5352	0.5046	0.5657
40	0.5981	0.6074	0.5775	0.6373
45	0.6502	0.6523	0.6232	0.6815
50	0.6969	0.7158	0.6882	0.7434
55	0.7385	0.7393	0.7124	0.7661
60	0.7753	0.7773	0.7519	0.8028
65	0.8076	0.8057	0.7814	0.8299
70	0.8358	0.8525	0.8308	0.8743

† Approximated power is based on the probability distribution method. Estimated power is based on the simulation method with 1024 simulated samples.
‡ 95% confidence limits for the estimated power.

One could also change the variance in the PARMS statement to find the largest variance for which four replications provide an 80% chance of detecting the treatment difference. Doing this, we find that the largest the variance can be in this case is $\sigma^2 = 0.18$. Finally, one could modify the variable mu to determine the minimum treatment difference four replications could detect at a significance level of $\alpha = 0.05$ and power = 0.80, with $\sigma^2 = 1$. For example, $mu = 10$ and 12.4 for $trt = 0$ and 1, respectively (a 24% difference), yields a power of 0.806.

The power calculations above are based on a generalized linear mixed model analysis, and therefore are based on an F statistic. The F distribution is an approximation to the true sampling distribution of the generalized linear mixed model test statistic. Hence, the power values obtained above are approximations as well. We can assess the accuracy of these approximations by estimating the true power using the simulation method discussed in Section 7.4. To do this, 1024 independent random samples were generated according to the model using the same combinations of assumed variance and number of blocks considered in the probability distribution method above. Each sample was analyzed using the GLIMMIX model shown in Fig. 7–4 (excluding the *parms* statement and the *noprofile* option). The results of the analyses of the simulated samples were used to calculate point and confidence interval estimates of the true power for each number of blocks under consideration. As can be seen in Tables 7–1, 7–2, and 7–3, in most cases the approximated power values are contained within the 95% confidence interval estimates of the true power. From this we can conclude that the power approximations obtained from the probability distribution method are accurate in this scenario. This illustrates the general result that for response variables with a normal distribution, this approximation is very good, and therefore we can be confident in the results provided by the probability distribution method in such cases. For response variables with non-normal distributions, using simulation to verify the results obtained from the probability distribution method is much more important because the non-normal case has not been studied as extensively and less is known about its performance in certain cases, such as when the number of replications is small. ∎

This simple example demonstrates the use of the probability distribution and simulation methods for evaluating power and precision. The remaining examples show how these methods can be used to perform power and precision analysis for several more realistic situations involving generalized linear mixed models.

7.6 A FACTORIAL EXPERIMENT WITH DIFFERENT DESIGN OPTIONS

The example in this section shows three alternative ways of setting up a two-factor factorial experiment with a given set of experimental units. Each design exhibits different power and precision characteristics, thereby providing the scientist with choices on ways to obtain more information from a fixed set of resources.

EXAMPLE 7.2

A researcher wants to conduct a field experiment to compare two treatments at three rates of application. For example, the two treatments could be two methods of application, two tillage methods, or two varieties. The three rates of application could represent amounts of a fertilizer or pesticide or irrigation levels. Treatment designs identical or similar to this two-treatment × three-rate factorial occur frequently in agronomic research. Assume that the response is normally distributed.

FIG. 7-9. Field layout of the experimental plots for a 3 × 2 factorial treatment structure for Examples 7.2, 7.5, 7.6, and 7.7.

Now suppose the resources available to the researcher consist of an 8 × 3 grid of plots with a gradient parallel to the direction of the 8-plot rows. Figure 7–9 shows the field layout.

The variation among the three-plot columns due to the gradient suggests that some form of blocking is advisable. Since there are six treatment × rate combinations in the treatment design, one obvious blocking strategy would combine pairs of adjacent columns into blocks, resulting in a randomized complete block (RCB) design with four blocks. However, with a strong enough gradient, adjacent columns may be dissimilar, resulting in excessively heterogeneous experimental units within blocks, a well-known poor design idea. An alternative design would use each 3-plot column as an incomplete block and set the experiment up as an incomplete block (IB) design with 8 blocks. A third approach would be to form blocks as in the randomized complete block design, assigning treatments to 3-plot columns within a block (whole plots), and then randomly assigning rates to sub-plots within each whole plot, resulting in a split plot (SP) design with an RCB whole plot design structure. Figure 7–10 shows a layout for each design.

Each design in Fig. 7–10 requires a different model for analysis, resulting in potentially different power characteristics. Each model consists of a component related to the treatment structure and a component related to the design structure of the experiment (Milliken and Johnson, 2009). The treatment structure is the same for all three experiments. Each model has in common the treatment × rate structure given by

$$\mu_{ij} = \beta_0 + T_i + D_j + TD_{ij}$$

where μ_{ij} is the mean for the ith treatment and jth rate, β_0 is the intercept, T_i is the ith treatment effect, D_j is the jth rate effect, and TD_{ij} is the effect of the treatment × rate interaction. The models differ in their blocking and error structures, which make up the remainder of each model. The complete models are given below.

- Randomized complete block (RCB):

$$Y_{ijk} = \mu_{ij} + R_k + e_{ijk}^R$$

where R_k is the kth block effect, assumed to be independent $N(0, \sigma_R^2)$, and e_{ijk}^R is the error term, assumed to be independent $N(0, \sigma_{eR}^2)$. The subscript and superscript R denotes the randomized complete block design.

FIG. 7-10. Field layouts as a randomized complete block, an incomplete block, and a split plot with whole plots in blocks for Example 7.2.

Randomized complete block

Block 1		Block 2		Block 3		Block 4	
T2 D1	T2 D3	T1 D3	T1 D2	T2 D1	T1 D1	T2 D1	T1 D2
T1 D3	T2 D2	T1 D1	T2 D3	T2 D2	T2 D3	T1 D3	T2 D3
T1 D2	T1 D1	T2 D1	T2 D2	T1 D3	T1 D2	T2 D2	T1 D1

Incomplete block

Block 1	Block 2	Block 3	Block 4	Block 5	Block 6	Block 7	Block 8
T2 D3	T2 D1	T2 D3	T2 D2	T2 D1	T2 D2	T2 D3	T2 D3
T1 D2	T1 D3	T1 D3	T2 D1	T1 D3	T1 D3	T2 D1	T2 D2
T1 D1	T1 D1	T1 D1	T1 D1	T1 D2	T1 D2	T1 D2	T1 D3

Split plot with whole plot in blocks

Block 1		Block 2		Block 3		Block 4	
T1 D1	T2 D2	T2 D2	T1 D3	T2 D3	T1 D1	T1 D2	T2 D3
T1 D3	T2 D1	T2 D3	T1 D2	T2 D1	T1 D2	T1 D1	T2 D2
T1 D2	T2 D3	T2 D1	T1 D1	T2 D2	T1 D3	T1 D3	T2 D1

- Incomplete block design (IB):

$$Y_{ijk} = \mu_{ij} + B_k + e^{I}_{ijk}$$

  where B_k is the kth incomplete block effect, assumed to be independent $N(0, \sigma_B^2)$, and e^{I}_{ijk} is the error term, assumed to be independent $N(0, \sigma_{eI}^2)$. The subscript and superscript I denotes the incomplete block design.

- Split plot (SP):

$$Y_{ijk} = \mu_{ij} + R_k + w_{ik} + s_{ijk}$$

  where R_k is the kth block effect (as in the RCB), w_{ik} is the whole plot error, assumed to be independent $N(0, \sigma_W^2)$, and s_{ijk} is the split plot error, assumed to be independent $N(0, \sigma_S^2)$.

Once we specify the plausible designs and their associated models, we have a decision to make. Which one should the researcher use? Assuming that the design costs are the same for the above designs, the answer is the design that maximizes power and precision for the treatment comparisons that address the research-er's objectives. To do the power analysis required to make this determination, we need to specify values of the μ_{ij} under the research hypothesis; that is, what are the "agronomically relevant" differences among these treatments and what comparisons among them best address the objectives?

As an example, suppose treatment 1 historically showed a 5.72 unit increase each time the rate was increased; e.g., from "low" to "medium" or from "medium" to "high." Suppose that the research hypothesis states that under treatment 2, the response to these rate increases would be greater. The researcher considered that a doubling of that rate would be "agronomically relevant." A power analysis for this research hypothesis is accomplished by performing a test of the equality of the linear effect of rate across each treatment, i.e., by testing the treatment × linear rate interaction. This hypothesis is tested in GLIMMIX using the following *contrast* statement:

contrast 'trt × lin_rate' trt*rate −1 0 1 1 0 −1;

FIG. 7-11. SAS statements to create an exemplary data set for the split plot design for Example 7.2.

```
data exemplary_data;
  input trt rate mu;
  do block=1 to 4 by 1;
    output;
  end;
datalines;
1 1 105.72
1 2 111.44
1 3 117.16
2 1 111.44
2 2 122.88
2 3 134.32
run;
```

Figure 7–11 shows the SAS data step used to create an exemplary data set for this power and precision analysis based on four blocks. The values for *mu* follow from the discussion above and the assumption that the mean response in the absence of any treatment is 100 units (any value could be used for this baseline).

Once the exemplary data set is specified, we need to specify the variance components associated with each design so that we can determine the non-centrality parameter. The variance structure in each model is a combination of the variance among plots within each column and the magnitude of the gradient. Suppose that enough is known about this structure to give the following information about the probable variance components that would result from each design.

- Randomized complete block (RCB): $\sigma_R^2 = 15$ and $\sigma_{eR}^2 = 34$
- Incomplete block design (IB): $\sigma_B^2 = 35$ and $\sigma_{eI}^2 = 14$
- Split plot (SP): $\sigma_R^2 = 15$, $\sigma_W^2 = 20$ and $\sigma_S^2 = 14$

Notice the difference between the RCB and IB variance components. The 3-plot columns are natural blocks induced by the gradient. The complete blocks are artificial "convenience" blocks constructed by combining natural blocks. Creating artificial blocks in this way reduces the variance among blocks and increases the error variance within blocks. This will affect power and precision.

Table 7–4 shows results from the precision analysis for these designs, specifically, the standard errors for various differences under each design. Note that the incomplete block design is best suited for comparisons between treatments (both main effects and simple effects at given rates), whereas the split plot design is best suited for comparisons among rates (split plot factor) but is least suited for comparisons among treatments (whole plot factor). For every effect, the randomized complete block design is less precise than the incomplete block.

TABLE 7–4. Precision analysis of competing designs for 3×2 factorial experiment in Example 7.2. Standard errors in bold indicate the best design for the corresponding effect.

	Design		
Effect	Randomized complete block	Incomplete block	Split plot
Treatment main effect	2.04	**1.80**	3.51
Rate main effect	2.50	1.98	**1.87**
Simple effect: treatment at jth rate	3.54	**2.99**	4.12
Simple effect: rate at ith treatment	3.54	2.86	**2.65**

FIG. 7–12. GLIMMIX statements to compute terms needed for the power/precision analysis for the split plot for Example 7.2.

```
proc glimmix data=exemplary_data noprofile;
   class block trt rate;
   model mu=trt rate trt*rate / dist=normal link=id;
   random intercept trt / subject=block;
   parms (15)(20)(14) / hold=1,2,3;
   contrast 'trt x lin_rate' trt*rate -1  0  1  1  0 -1;
   lsmeans trt | rate / diff slicediff=(trt rate);
   ods output contrasts=research_H_test_terms;
run;
```

Figure 7–12 shows the GLIMMIX statements needed to obtain the values for a power analysis for the split plot version of the experiment. The *random* statement accounts for the design structure of the experiment by incorporating random effects for the block and whole plot error terms. The *contrast* statement tests the research hypothesis of interest and produces values needed for the power analysis. The *ods* statement saves these values in a dataset we have called *research_h_test_terms*. These values are then processed as shown in Fig. 7–7 of Example 7.1. More than one *contrast* statement can be included and the *ods* statement can create multiple output data sets; for example, if one also wanted to output the type 3 test of fixed effects (*tests3*) results. The *lsmeans* statement produces results for the precision analysis.

By changing the *random* statements different design structures can be accommodated. This was done to calculate the power for the RCB and IB versions of the experiment as well. The power approximations for the three designs obtained using the probability distribution method are given below:

- Split plot (SP): approximated power 0.801
- Incomplete block (IB): approximated power 0.726

- Randomized complete block (RCB): approximated power 0.451

The three designs provide different levels of power. These results underline the take-home message of this example; namely, over-simplification of power analysis and sample size analysis often encourages misplaced focus in designing an experiment. To emphasize this point, consider the following scenario. Imagine that the researcher has done all the planning up to the point where the power is actually computed. The researcher, having had only a semester of statistical methods, is familiar only with randomized complete block designs and, therefore, has considered only that design. Shortly before the grant proposal is to be submitted, the researcher brings the statistician the information about the variance components and the agronomically relevant difference and asks for a power calculation, using the standard greeting, "I know you're busy, but I need this by noon today."

Once the power is computed, the statistician delivers the bad news. The power for four blocks is only 0.45. "How many blocks do I need to get the power up to 0.80?" By running the power algorithm above with different numbers of blocks, the statistician finds that nine blocks would be required. The researcher adjusts the budget to accommodate nine blocks and everyone lives happily ever after—except those whose money and labor have been wasted. The researcher has asked the wrong question. Rather than "How many blocks do I need?" the question should have been "What is the most efficient way to use the resources that I have available?" And, the researcher should also have asked this question much sooner. This conversation should have begun when the researcher was first thinking about this project. This scenario illustrates a point that should have particular resonance in a time of budget deficits, unpredictable energy costs, and tight money. ∎

7.7 A MULTI-LOCATION EXPERIMENT WITH A BINOMIAL RESPONSE VARIABLE

This section illustrates another common experimental setting. From the statistical perspective, multi-location studies present the same basic statistical issues as laboratory studies conducted in multiple growth chambers or using other types of "identical" equipment or in studies conducted in multiple independent runs over time. In addition, some of the issues involved in designing experiments where the response of interest is a proportion are discussed. The considerations in these examples are applicable to any binomial response variable—dead/alive, damaged/undamaged, germinate/did not germinate, etc. There are standard textbook formulas for determining sample size with binomial response variables. However, as the examples will show, the standard formulas are inappropriate and inapplicable to the vast majority of agronomic experiments in which conclusions are to be based on binomial response variables. The examples demonstrate an alternative that is applicable to these types of experiments.

EXAMPLE 7.3

In this example, the objective is to compare the effects of two treatments on the proportion of surviving plants when exposed to a certain disease. Suppose that a standard treatment is to be compared to a new experimental treatment, and that experience with the standard treatment suggests that the proportion of plants exposed to the disease that survive averages 15%. It is believed that the experimental treatment can increase that proportion to 25%. The researcher wants to know how many plants per treatment must be observed to have a reasonable chance of detecting such a change.

Some experimental design textbooks have tables giving the needed sample size based on standard formulas for binomial response variables (e.g., Cochran and Cox, 1992). Alternatively, one could use standard power and sample size software, such as PROC POWER in SAS. Either approach yields a required sample size of 250 plants per treatment to have power of 0.80 when a significance level of $\alpha = 0.05$ is used. The GLIMMIX based probability distribution approach would yield the same answer if one uses the program shown in Fig. 7–13. This program assumes a binomial generalized linear model with a logit link. The model is given by

$$\text{logit}(\pi_i) = \beta_0 + T_i$$

FIG. 7-13. GLIMMIX statements to obtain the power for a binomial response for Example 7.3.

```
data binomial;
   input trt n p;
   expected_y = n*p;
   datalines;
   0  250  0.15
   1  250  0.25
run;

proc glimmix data=binomial initglm;
   class trt;
   model expected_y/n = trt / chisq dist=bin link=logit;
   ods output tests3=power_terms;
run;

data power;
   set power_terms;
   alpha = 0.05;
   non_cent_parm = numdf*chisq;
   chi_sq_critical = Cinv(1 - alpha, numdf, 0);
   power = 1 - ProbChi(chi_sq_critical, numdf, non_cent_parm);
run;
```

where π_i is the probability that a plant survives when the ith treatment is applied, β_0 is the intercept and T_i is the ith treatment effect. Note that this model is a true generalized linear model and, hence, uses a χ^2 statistic to test the equality of the π_i. The *chisq* option on the *model* statement requests the χ^2 test. Figure 7–13 also shows the statements needed to compute the power for this model. Note that these statements take into account the fact that the χ^2 distribution is being used as the basis for inference for this model.

Unfortunately, this approach is overly simplistic and misleading for most agronomic research. Most agronomic experiments involve some form of blocking and are often conducted at multiple locations. To see how this affects power, suppose that the proposed experiment is to be performed at four locations. The researcher asks, "If I need 250 plants per treatment, should I divide them equally among the four locations?"

A model that reflects this design is given by

$$\text{logit}(\pi_{ij} \mid L_j, TL_{ij}) = \beta_0 + T_i + L_j + TL_{ij}$$

where π_{ij} is the probability that a plant survives when the ith treatment is applied at the jth location, T_i is the ith treatment effect, L_j is the jth location effect, and TL_{ij} is the treatment $\times$ location interaction effect. If locations represent a random sample from the target population, then location and treatment $\times$ location are random effects, where L_j are independent $N(0, \sigma_L^2)$, TL_{ij} are independent $N(0, \sigma_{TL}^2)$, and the L_j and TL_{ij} are assumed to be independent.

It is important to understand what the variance components for location and treatment $\times$ location signify because they are critical to getting the design correct for this experiment. In categorical data, the ratio $\pi/(1 - \pi)$ represents the odds of the event of interest. The logit of π is the natural logarithm of these odds. The odds ratio is defined to be the odds for the experimental treatment divided by the odds for the reference treatment. The difference between the logits for the two treatments is the log odds ratio. Therefore the variance component σ_L^2 measures the variation in the log odds from location to location averaged over treatments and σ_{TL}^2 measures the variation in the log odds ratio among treatments from location to location. For example, if the probability of a plant surviving averages 0.15 for the reference treatment (and as a result, the log-odds of survival averages −1.73), the actual probability varies from location to location and between treatments over locations. With a little reflection this makes sense because the motivation for multi-site experiments is the implicit assumption that variation exists among locations and one wants to avoid experimental results that are site-specific.

How can one anticipate values of σ_L^2 and σ_{TL}^2 for power or precision analysis and planning experiments? Historical data could provide guidance. Otherwise, the researcher could "guesstimate" the lowest and highest values of π likely to occur among the locations in the population. For example, suppose, based on historical data a researcher "guesstimates" that for the reference treatment $\pi = 0.1$ is the minimum probability of a plant surviving considered plausible at any give location and that $\pi = 0.2$ is the maximum. Converting from the data scale to the model scale, the plausible range of logits across locations is −2.20 to −1.39. The standard deviation can then be approximated as the difference between the maximum and the minimum divided by six, or roughly 0.135. Hence, the variance among logits is approximately $(0.135)^2 = 0.018$. This can serve as an approximation for σ_L^2. If similar variation occurs for the experimental treatment, then odds ratios could vary from 1.0 (when $\pi = 0.2$ for both the reference and experimental treatments) to 3.86 (when $\pi = 0.1$ for the reference treatment and $\pi = 0.3$ for the experimental treatment). The log odds ratio would then vary from 0 to 1.35, yielding a variance of $(0.135/6)^2 = 0.05$ as an approximation for σ_{TL}^2. In this way approximate values for the variances of the location and treatment × location random effects can be obtained.

For this example, round off the approximate variance components obtained above; that is, use "best guesses" of $\sigma_L^2 = 0.02$ and $\sigma_{TL}^2 = 0.05$, respectively. Suppose the researcher proposes to observe 65 plants per treatment at each of the four locations. Figure 7–14 shows the SAS statements needed to approximate the power using the probability distribution method.

FIG. 7-14. SAS program to determine the approximated power for the multi-location binomial experiment in Example 7.3.

```
data multi_loc_binomial;
    input trt n pi;
    do location = 1 to 4 by 1;
        expected_y = n*pi;
        output;
    end;
    datalines;
    0 65  0.15
    1 65  0.25
    run;

proc glimmix data=multi_loc_binomial;
    class location trt;
    model expected_y/n = trt / dist=bin link=logit;
    random intercept trt / subject=location;
    parms (0.02) (0.05) / hold=1,2;
    ods output tests3=power_terms;
    run;

data power;
    set power_terms;
    alpha = 0.05;
    non_cent_parm = numdf*Fvalue;
    F_critical = Finv(1 − alpha, numdf, dendf, 0);
    Power = 1 − ProbF(F_critical, numdf, dendf, non_cent_parm);
    run;
```

TABLE 7-5. Approximated and estimated power for 65 plants per location–treatment combination for Example 7.3.

Number of locations	Approximated power†	Estimated power†	Lower confidence limit‡	Upper confidence limit‡
4	0.36	0.277	0.257	0.296
8	0.80	0.838	0.822	0.854

† Approximated power is based on the probability distribution method. Estimated power is based on the simulation method with 2048 simulated samples.
‡ 95% confidence limits for the estimated power.

Since this is a generalized linear mixed model, the test for no treatment effect on the logit scale uses an F statistic. As a result, the subsequent computations necessary to calculate the power are exactly as shown previously in Fig. 7–7. This approach yields a power of 0.36, far less than the power of 0.80 often used in sample size calculations. The reason for the discrepancy is that standard power computations for binomial responses do not account for the variance among locations and, as a result, are vulnerable to dramatically overstating the power and understating the actual sample size requirements. One can vary the number of plants per location by changing n and vary the number of locations by changing the *do* statement to examine various design alternatives. With 65 plants per treatment group at each location, we see that eight locations are required to achieve power of at least 0.80 given the assumed variance components (Table 7–5).

Since the response variable is not normally distributed, this is a situation where it is important to use the simulation method to check of the accuracy of the probability distribution method. For various values for the number of sites and the total number of plants, 2048 independent samples were generated according to the model above. Each sample was analyzed using the GLIMMIX model shown in Fig. 7–14 after omitting the *parms* statement. The results obtained were then used to estimate the true power for each combination of the simulation parameters.

For example, both approximated and estimated power values that result by using four and eight sites with 65 plants per treatment group at each site are shown in Table 7–5. Note that the approximated power obtained from the probability distribution method using the GLIMMIX statements in Fig. 7–14 is higher than the power estimate obtained using the simulation method when four locations are used, but that the estimated power obtained using the simulation method is greater than the approximated power obtained from the probability distribution methods when eight locations are used. In other words, the probability distribution method gives a somewhat optimistic power approximation when the experiment is under-powered and a slightly pessimistic approximation when the experiment is adequately powered. Discrepancies aside, both the simulation and probability distribution power analyses give accurate assessments of whether the proposed number of locations is sufficient or not.

Table 7–6 gives results for additional combinations of number of locations and number of plants per treatment per location. Note that the total number of plants

TABLE 7-6. Approximated and estimated power for various numbers of locations and plants per location–treatment combination for Example 7.3.

Number of locations	Plants per location–treatment combination	Total number of plants per treatment	Approximated power†	Estimated power	Lower confidence limit‡	Upper confidence limit‡
10	26	260	0.63	0.622	0.601	0.643
10	43	430	0.80	0.823	0.806	0.839
20	13	260	0.72	0.742	0.723	0.761
20	15	300	0.78	0.771	0.753	0.790
20	16	320	0.80	0.799	0.782	0.817
50	6	300	0.83	0.833	0.817	0.850
132	2	264	0.80	0.811	0.794	0.828

† Approximated power is based on the probability distribution method. Estimated power is based on the simulation method with 2048 simulated samples.
‡ 95% confidence limits for the estimated power.

required decreases as the number of locations increases, but at no point is it possible to obtain 80% power with only 260 plants. Some researchers believe that the algorithm used by GLIMMIX does not produce accurate results when the "cluster size" (i.e., number of plants per location) is small. We observe this to be true for underpowered experiments (e.g., 2 plants per location and few locations), but not when the number of locations is sufficient for adequate power. This underlines the need to design experiments tailored to the distribution of the response variable to be analyzed and not to depend on conventional wisdom. ∎

EXAMPLE 7.4

As a variation on Example 7.3 that clearly illustrates the effect of the number of locations on power, suppose there are a total of 600 plants available and that they are to be divided equally between treatments among a number of locations to be used in the experiment. As in the previous example suppose that σ_L^2 = 0.02 and σ_{TL}^2 = 0.05. Using these assumed values for the variance components, what power can be achieved for detecting the difference between the proportions 0.15 and 0.25? Does the power depend on how many locations we use? If so, in what way does it matter?

Table 7–7 shows the power for this test as a function of the number of locations used from 2 to a maximum of 150. There are several things to notice from this analysis. First, we see that across the entire range of the number of locations we could use, the power increases as the number of locations increases. As might be expected, the per-location increase in power is greatest when the number of locations is small. The maximum power attainable is 0.85, which occurs when we use 150 locations. It appears that using between 25 and 30 locations results in a

TABLE 7-7. Approximated and estimated power for 600 total plants for Example 7.4.

Number of locations	Plants per Location	Approximated power†	Estimated power†	Lower confidence limit‡	Upper confidence limit‡	Number of samples†
2	300	0.1302	0.0000	0.0000	0.0000	2048
3	200	0.2640	0.0870	0.0747	0.0992	2047
4	150	0.3887	0.2954	0.2756	0.3151	2045
5	120	0.4818	0.4628	0.4412	0.4844	2044
6	100	0.5486	0.5340	0.5124	0.5556	2043
10	60	0.6844	0.6663	0.6459	0.6868	2041
12	50	0.7170	0.7383	0.7193	0.7574	2037
15	40	0.7485	0.7515	0.7327	0.7703	2024
20	30	0.7787	0.7756	0.7575	0.7938	2028
25	24	0.7960	0.7986	0.7812	0.8161	2031
30	20	0.8072	0.8286	0.8122	0.8451	2025
50	12	0.8288	0.8454	0.8295	0.8612	2011
60	10	0.8340	0.8321	0.8158	0.8484	2025
75	8	0.8391	0.8437	0.8279	0.8595	2028
100	6	0.8441	0.8559	0.8405	0.8712	2019
150	4	0.8491	0.8558	0.8405	0.8711	2018

† Approximated power is based on the probability distribution method. Estimated power is based on the simulation method. 2048 samples were simulated for each number of locations. The number of samples for which the GLIMMIX procedure converged successfully is given in the rightmost column. Section 2.7 briefly discusses the computational issues involved with convergence of the numerical algorithms used.

‡ 95% confidence limits for the estimated power.

power of 0.80. In addition, there is little reason to use more than 30 locations, since beyond this point the per-location increase in power is very low.

This is a situation where it is important to use simulation to verify the power approximations provided by the probability distribution method. Again, 2048 independent samples were generated for each combination of the number of locations and the number of plants per treatment per location. As can be seen in Table 7–7, in several cases the approximated power obtained from the probability distribution method is actually larger than the upper endpoint of the 95% confidence interval estimate of the power obtained from the simulation method. In particular, for this situation the values of the approximated power appear to be too large when fewer than five locations are considered. The power approximations provided by the probability distribution method appear to be most accurate when the number of locations is large, even though in those cases the number of plants at each location is small. This is a somewhat surprising result. The downside is that the proportion of samples for which the GLIMMIX estimation procedure converges tends to decrease as the number of plants per location decreases. ∎

7.8 A SPLIT PLOT REVISITED WITH A COUNT AS THE RESPONSE VARIABLE

In Example 7.2 the response variable was assumed to be continuous and normally distributed and the focus of the inference was the treatment × linear rate effect. Generalized linear mixed models were used to evaluate the power profiles of three potential designs (randomized complete block, incomplete block, and split plot) for the experiment. What if the response does not have a normal distribution? The approach presented in that example can be used to evaluate the power profile of one or more designs for other types of responses. In this section we show how this can be accomplished when the response of interest is a count (e.g., number of weeds or insects).

The probability distribution of counts in biological settings has received considerable attention in recent years. Young and Young (1998) provided a good summary of the main issues. Historically the Poisson has been the presumptive distribution for counts. One important characteristic of the Poisson distribution is that the mean and variance of the distribution are equal. This is a very strong assumption, and there is now considerable empirical evidence suggesting that biological count data that satisfy the Poisson assumption are very much the exception (Young and Young, 1998). On the other hand, much evidence supporting the use of other distributions, such as the negative binomial (Section 2.3), has accumulated from field studies over the past several decades.

The motivation for using the negative binomial rather than the Poisson is over-dispersion. Relative to the Poisson distribution, over-dispersion occurs whenever the variance is larger than the mean. It occurs with count data when biological entities (e.g., weeds, insects, mold, viruses) tend to cluster rather than disperse completely at random. The negative binomial distribution can account for events occurring at random with clustering, whereas the Poisson assumes events occurring completely at random. Hence, the negative binomial tends to be a better model for biological counts in many situations, and planning research under the Poisson assumption can result in serious, even disastrous mistakes in assessing sample size requirements.

In this section we focus on the negative binomial distribution. In addition, because count data are often analyzed using a normal approximation with transformations, typically the natural logarithm or square root of the counts, the implications of power analysis from the transformation perspective are also considered.

EXAMPLE 7.5

Generalized linear mixed models for count data typically use the natural logarithm as the link function. For the factorial treatment structure in Example 7.2, when the response is a count that is assumed to have a negative binomial distribution, the conditional model for the split plot design with whole plots in blocks can be written as

$$\log(\mu_{ijk} \mid R_k, w_{ik}) = \mu_{ij} + R_k + w_{ik}$$

where μ_{ijk} is the mean count for the ith treatment and jth rate in the kth block, μ_{ij} is the mean count for the ith treatment and jth rate, R_k is the kth block effect, assumed to be independent $N(0, \sigma_R^2)$, w_{ik} is the whole plot error, assumed to be independent $N(0, \sigma_W^2)$, and R_k and w_{ik} are assumed to be independent.

The variance component approximations required for a power analysis involving count data can be obtained using an approach similar to that used in Example 7.3 for a binomial response. One begins by determining the variability among counts, from the minimum to the maximum plausible among blocks and among whole plot experimental units for a given treatment × rate combination. Since the generalized linear mixed model models log-counts, we convert the minimum and maximum counts from the data scale (counts) to the model scale (log counts). The range on the log scale divided by six gives an approximation of the standard deviation, which when squared yields the approximate variance. In a split plot this procedure must be used to approximate the block variance as well as the whole plot variance.

If the mean of the negative binomial is denoted by μ then the variance is given by $\mu + k\mu^2$, where k is the scale or aggregation parameter ($k = 1/\delta$ in Table 2–2). The scale parameter must be positive. The negative binomial distribution is flexible in that the degree to which the variance exceeds the mean is allowed to vary. In particular, for a fixed value of the mean, the variance varies directly with the value of the aggregation parameter. For values of k close to zero, there is little over-dispersion and the variance is close to the mean, as in the Poisson distribution. The over-dispersion increases as k increases. In specifying a value of k for power and precision analysis using GLIMMIX, one chooses a value of k that reasonably approximates the anticipated mean–variance relationship.

One way to obtain a reasonable value of k is as follows. Identify the treatment conditions under which the researcher is most familiar with the distribution of counts. For example, in an experiment where an experimental treatment is being compared to a standard treatment, the researcher may be familiar with the distribution of counts under the standard treatment. The researcher can then identify the count that would be expected (μ) under that treatment, as well as the largest and smallest counts that would likely be expected under that treatment. Then an approximate value of k can be obtained from

$$k \cong \frac{\left[(\max - \min)/6\right]^2 - \mu}{\mu^2}$$

where max is the largest expected count and min is the smallest expected count under that treatment.

This technique requires the same kind of information regarding the variability of the response as in previous examples and should give a reasonable value of k to use in the calculations.

Suppose that the block variance has been determined to be approximately 0.25 and the whole plot variance approximately 0.15. In addition, the researcher has indicated that when the expected count is 10, then about 50 would be the largest count and 4 the smallest count they would expect to see. With these values, an approximate value for the scale parameter is

$$k \cong \frac{\left[(50-4)/6\right]^2 - 10}{10^2} = 0.49$$

which will be rounded off to $k = 0.5$.

As before, the focus is on inference about the treatment $\times$ linear rate effect. Suppose that the researcher is interested in detecting a difference in the linear rate effect when it is three times higher under treatment 2 than it is under treatment 1. In addition, she is interested in determining the number of blocks required to have 80% power of detecting such a difference.

Figure 7–15 shows the SAS statements to create an exemplary data set for this analysis when four blocks are used. The response variable is labeled *expected_count*. Figure 7–16 shows the GLIMMIX statements that provide the values needed to obtain the non-centrality parameter and the degrees of freedom for the power analysis. The *initglm* option on the *proc* statement instructs GLIMMIX to use generalized linear model estimates as initial values for fitting the generalized linear mixed model. The first two terms in the *parms* statement are the block and whole plot variance estimates, respectively. The third term is the aggregation parameter k. While Fig. 7–15 and 7–16 are for a split plot design, they can be modified to accommodate other design structures such as the randomized complete block and incomplete

FIG. 7–15. SAS statements to create an exemplary data set for Example 7.5.

```
Data Split_Plot_with_Counts;
    input trt rate expected_count;
    do block=1 to 4 by 1;
        output;
    end;
    datalines;
1  1  10
1  2  9
1  3  8
2  1  9
2  2  6
2  3  3
run;
```

FIG. 7-16. GLIMMIX statements for the power analysis for negative binomial model for Example 7.5.

```
proc glimmix data=Split_Plot_with_Counts initglm;
    class block trt rate;
    model expected_count=trt rate trt*rate / dist=NegBin link=log;
    contrast 'trt x lin rate' trt*rate −1 0 1 1 0 −1;
    lsmeans trt|rate / diff slicediff=(trt rate) ilink;
    random intercept trt / subject=block;
    parms (0.25)(0.15)(0.50) / hold=1, 2, 3;
    ods output contrasts=power_terms;
run;
```

TABLE 7-8. Approximated and estimated power for the split plot design with the negative binomial distribution in Example 7.5.

Number of blocks	Approximated power†	Estimated power†	Lower confidence limit‡	Upper confidence limit‡	Number of samples†
4	0.1670	0.1906	0.1662	0.2150	997
10	0.3782	0.4011	0.3689	0.4333	890
20	0.6576	0.6410	0.6100	0.6720	922
27	0.7877	0.8000	0.7744	0.8256	940
28	0.8024	0.7871	0.7611	0.8130	958

† Approximated power is based on the probability distribution method. Estimated power is based on the simulation method. 1024 samples were simulated for each number of blocks. The number of samples for which the GLIMMIX procedure converged successfully is given in the rightmost column. Section 2.7 briefly discusses the computational issues involved with convergence of the numerical algorithms used.
‡ 95% confidence limits for the estimated power.

block alternatives discussed previously. This can be done regardless of the distribution assumed for the counts. As in the case of the normal distribution, the block variance changes depending on the proposed design. That is, if natural blocks of size three are combined into complete but heterogeneous blocks of size six, block to block variability will necessarily decrease as within block (whole plot) variability increases. Increasing within block heterogeneity will also increase over-dispersion.

The power associated with different numbers of blocks can be obtained by varying the upper bound in the *do* statement in Fig. 7–15. The results for various numbers of blocks are given in Table 7–8 for the split plot. With four blocks, there is only approximately a 17% chance of detecting a threefold difference in linear rate effects. To achieve 80% power, 28 blocks would be needed. ∎

Two questions arise at this point. First, if we assume a Poisson distribution for the counts, will the results change? If so, how? Second, what if the power analysis is based on a normal approximation using a transformation such as the logarithm or square root of the counts? These questions are considered in the following examples.

EXAMPLE 7.6

This example is a continuation of Example 7.5 for the split plot design in which the response is assumed to follow a Poisson distribution. Since the Poisson generalized linear mixed model is also on the log scale, the process that led us to assuming block and whole plot variances of 0.25 and 0.15, respectively, for the negative binomial would lead us to the same anticipated variance components for the Poisson. However, estimation of the scale parameter k in the negative binomial would not be applicable. If one computes the approximated power under the Poisson assumption, a power of 50% for four blocks is obtained. Only eight blocks are needed to obtain a power over 80%. The power for this situation, accounting for over-dispersion using the negative binomial distribution, would only be 31%. Failing to account for over-dispersion by assuming a Poisson distribution generally results in severely underestimating the resources needed for adequate power. ∎

EXAMPLE 7.7

This example is a continuation of Example 7.5 for the split plot design in which the transformed counts are assumed to be approximately normally distributed. To assess the power using the methods in Example 7.2 following a transformation such as the logarithm or square root of the count, we would use the same exemplary data set as shown in Fig. 7–15. As with the negative binomial power assessment (Fig. 7–16) we would need to determine the approximate variance components. If the log transformation were used, the variance components for block and whole plot error would be the same as for the generalized linear mixed model with log link. If the square root transformation were used, the variances from the log scale would need to be rescaled to the square root scale. Only the log scale will be considered in detail here. While not shown, the square root transformation produced similar results.

If the normal approximation is used, an estimate of the split plot error variance is required in addition to the block and whole plot variance components. This is where the problem with using the normal approximation to assess power occurs. Assume that as before, the expected smallest count is 4 and the expected largest is 50. Then we could anticipate the split plot variance to be approximately

$$\left[\frac{\log(50) - \log(4)}{6} \right]^2 = 0.177$$

Alternatively, the formula for the variance of the negative binomial could be used to produce an estimate which is then transformed to the log scale. For $k = 0.5$ and $\mu = 10$, the split plot variance of the counts would be $\mu + k\mu^2 = 10 + 0.5(10)^2 = 60$. This is important because it might very well be the variance of counts that appears in literature reviews of similar experiments that are often the source of the variance information in power analyses. Using the delta method (Section 3.2), if the variance on the count scale is 60, the variance on the log scale is given by

$$\left[\frac{\partial \log(\mu)}{\partial \mu}\right]^{2} \text{var}(count) = \left(\frac{1}{\mu}\right)^{2} \text{var}(count) = \left(\frac{1}{10}\right)^{2} 60 = 0.60$$

The GLIMMIX statements shown in Fig. 7–17 can be used to assess the power assuming that the estimated split plot variance is 0.177. However, the results will be quite different if the estimated split plot variance is 0.60. Two defensible approaches in this case lead to different variances. Which should be used? There is no clear answer.

For the normal approximation assuming a split-plot variance of 0.177, the resulting power for four blocks is 48.2% (not shown). For 28 blocks (the required number of blocks assuming the negative binomial), the power is greater than 99.9%. Eight blocks are required to obtain power of at least 80%. This result is similar to what would be obtained with the Poisson distribution. On the other hand, if the power analysis is based on a split-plot variance of 0.60, the power for four blocks is 18.1%, for 28 blocks it is 84.2%, and the required number of blocks for 80% power is 26. All of this assumes that the log counts have an approximately normal distribution.

These results suggest two things. First, using the normal approximation, very different variance estimates and, hence, very different power assessments can be obtained. Using the crude approximation of variance,

$$\left[\frac{\log(\max) - \log(\min)}{6}\right]^{2}$$

where max is the highest plausible count and min is the lowest plausible count, can result in a very optimistic split plot variance and, hence, a power assessment as misleading as the one based on the Poisson distribution. On the other hand, if the variance from the negative binomial is transformed to the log scale for use

FIG. 7–17. GLIMMIX statements for power analysis for log counts assumed to be approximately normally distributed for Example 7.7.

```
proc glimmix data=Split_Plot_with_Counts initglm;
    class block trt rate;
    log_count = log(expected_count);
    model log_count = trt rate trt*rate;
    contrast 'trt x lin rate' trt*rate -1 0 1 1 0 -1;
    lsmeans trt|rate / diff slicediff=(trt rate) ilink;
    random intercept trt / subject=block;
    parms (0.25)(0.15)(0.177) / hold=1,2,3;
    ods output contrasts=power_terms;
run;
```

with the log count normal approximation, the resulting power calculations are quite different. In this case, they were quite similar to results obtained with the negative binomial, but there is no guarantee what will happen in general.

The second conclusion about the normal approximation follows from the first. There are multiple plausible ways to approximate the variance and, as illustrated above, these include an approach that seems perfectly reasonable but gives a disastrously optimistic assessment of power. Therefore, we suggest avoiding the normal approximation as a tool for power analysis with count data. ∎

There are two overriding take-home messages for handling count data. First, do not use the Poisson distribution to plan experiments for count data. There will almost certainly be some level of over-dispersion present in count data. As has been demonstrated above, use of the Poisson distribution can drastically underestimate the number of replications needed. Even in those situations where the amount of over-dispersion is very small and the Poisson may give acceptable results, the negative binomial distribution can still be used since small values of over-dispersion can be accounted for through a small value of the scale parameter k. Now that software is available that can fit the negative binomial distribution, we recommend that it be used in planning experiments involving count data. There appears to be no compelling reason to use the Poisson distribution, at least at the planning stages of such an experiment.

Second, planning experiments where count data are the primary responses requires good preliminary information about the anticipated variability. Power computations are sensitive to the scale parameter when using the negative binomial and to the error variance when using the normal approximation. Without good preliminary information, one risks potentially serious errors in planning. One may either overestimate power and hence understate the amount of needed replication or underestimate power and hence overstate the number of replications needed. Obviously, the goal is to avoid either case.

7.9 SUMMARY AND CONCLUSIONS

The generalized linear mixed model based methods to assess power and precision can be applied to any proposed design for which a generalized linear mixed model will be used to analyze the data. The following information is needed:

- the anticipated distribution of the response variable,
- the anticipated magnitude of the variance components implied by the proposed design—for non-normal generalized linear mixed models, care must be taken to express the variance components on the model scale and not the data scale,
- the objectives expressed as testable hypotheses (for power analysis) or interval estimates (for precision analysis),
- for power analysis, the minimum scientifically relevant magnitude of the effect to be detected.

One of the benefits of this approach to power and precision analysis is the requirement that an exemplary data set must be created and GLIMMIX statements to analyze that data set must be written to obtain the needed terms for the power analysis. This is essentially a dress-rehearsal for actual analysis once the data are collected. Subsequently, the researcher is less likely to think, "Now what?" once the data are collected and ready to be analyzed.

Generalized linear mixed model based power or precision analysis also encourages, or should encourage, an early conversation between the researcher and the statistical scientist. As Examples 7.2 and 7.3 clearly illustrate, the terms power analysis and sample size determination often lead researchers to misunderstand the point. Sample size requirements for a badly conceived design can be needlessly high. There are frequently much more efficient designs that researchers cannot be expected to know about, but statistical scientists, given adequate information, can easily suggest. The real question is how to use experimental resources most efficiently, which absolutely mandates involving the statistical scientist in the discussion much earlier than is unfortunately common practice in far too many cases. In an era of tight budgets, this point cannot be emphasized too forcefully.

Finally, the generalized linear mixed model based probability distribution method, in knowledgeable hands, offers a quick way to consider plausible design alternatives. The caveat is that because these methods are relatively new and knowledge about their behavior, especially at the margins, is an active area of research in statistics, the final design choices should be verified via simulation to reduce the chances of unpleasant surprises once the data are collected.

REFERENCES CITED

Cochran, W.G., and G.M. Cox. 1992. Experimental designs. 2nd ed. John Wiley and Sons, New York.

Hahn, G.J. 1984. Experimental design in the complex world. Technometrics 26:19–31. doi:10.2307/1268412

Hinkelmann, K., and O. Kempthorne. 1994. Design and analysis of experiments. Vol. I. Introduction to experimental design. John Wiley and Sons, New York.

Light, R.J., J.D. Singer, and J.B. Willett. 1990. By design: Planning research on higher education. Harvard Univ. Press, Cambridge, MA.

Littell, R.C. 1980. Examples of GLM applications. p. 208–214. *In* Proceedings of the fifth annual SAS Users Group International conference. SAS Institute, Cary, NC.

Littell, R.C., G.A. Milliken, W.W. Stroup, R.D. Wolfinger, and O. Schabenberger. 2006. SAS for mixed models. 2nd ed. SAS Institute, Cary, NC.

Lohr, V.I., and R.G. O'Brien. 1984. Power analysis for univariate linear models: SAS makes it easy. p. 847–852. *In* Proceedings of the ninth annual SAS Users Group International conference. SAS Institute, Cary, NC.

Mead, R. 1988. The design of experiments: Statistical principles for practical applications. Cambridge Univ. Press, Cambridge, UK.

Milliken, G.A., and D.E. Johnson. 2009. Analysis of messy data. Volume I: Designed experiments. 2nd ed. CRC Press, Boca Raton, FL.

O'Brien, R.G., and V.I. Lohr. 1984. Power analysis for linear models: The time has come. p. 840–846. *In* Proceedings of the ninth annual SAS Users Group International conference. SAS Institute, Cary, NC.

Stroup, W.W. 1999. Mixed model procedures to assess power, precision, and sample size in the design of experiments. p. 15–24. *In* Proceedings of the 1999 Biopharmaceutical Section, American Statistical Association. American Statistical Assoc., Alexendria, VA.

Stroup, W.W. 2002. Power analysis based on spatial effects mixed models: A tool for comparing design and analysis strategies in the presence of spatial variability. J. Agric. Biol. Environ. Stat. 7:491–511. doi:10.1198/108571102780

Young, L.J., and J.H. Young. 1998. Statistical ecology: A population perspective. Kluwer Academic Publishers, Norwell, MA.

CHAPTER 8

PARTING THOUGHTS AND FUTURE DIRECTIONS

8.1 THE OLD STANDARD STATISTICAL PRACTICE

In considering the range of material and examples covered in this book, one overriding reality should stand out; namely, the generalized linear mixed model (GLMM) has substantially changed what is regarded as "standard statistical practice" relative to what was regarded as standard practice even as recently as a decade ago. It is instructive at this point to revisit Table 1–1 from the perspective of one asking "What is standard statistical practice?" As noted in Chapter 1, for much of the 20th century and even into the beginning of the 21st century, analysis of variance (ANOVA) and regression have been the dominant tools of statistical practice in experimental research, including research in the agricultural and natural resources sciences. As recently as the 1990s, if you had asked a statistical scientist what was meant by linear models, the answer would have been ANOVA and regression type analyses assuming normally distributed data with independent errors and equal variances.

Times have changed. Quantum increases in computing capacity have rapidly enabled advances in statistical theory to be put into practice. Table 1–1 makes it clear that contemporary researchers deal with observations on variables whose types cover the full range from continuous and categorical responses to counts and time to event measurements with a matching breadth of probability distributions. The complexity of experiments conducted in modern research makes demands for analyses over the full range of explanatory and random model components listed across the table. In this book we have seen that generalized linear mixed models, unlike the rather restrictive "general" linear model that dominated the 20th century, move rather easily among most of the response variable-by-model combinations in the table. Examples covering most of these scenarios have been presented in the previous chapters.

The types of response variables and modeling issues portrayed in Table 1–1 were present long before GLMM theory was developed and modern computers made their implementation practical. So what did standard statistical practice

doi:10.2134/2012.generalized-linear-mixed-models.c8

Analysis of Generalized Linear Mixed Models in the Agricultural and Natural Resources Sciences
Edward E. Gbur, Walter W. Stroup, Kevin S. McCarter, Susan Durham, Linda J. Young, Mary Christman, Mark West, and Matthew Kramer

mean in the pre-GLMM era? The dominant theory, epitomized by software such as PROC GLM in SAS (SAS Institute, Cary, NC), assumed linear models containing only fixed effects, with independent, normally distributed observations and homogeneous variances. Non-normality, heterogeneous variances, and lack of independence each represented a data analysis crisis requiring immediate attention. Standard statistical practice came to include standard fixes to make the data suitable for normal linear models with the standard assumptions. For random effects, linear model software, whose internal architecture was based on fixed effects only linear model theory, was equipped with options to compute expected mean squares, use non-default error terms, and compute p-value adjustments for certain kinds of repeated measures data that violated independence assumptions—options that in retrospect are little more than partially effective "band-aid" solutions. But, as we have seen, for example in the split plot examples in Chapter 4, the band-aids were never enough. The eventual replacement of software such as PROC GLM with true mixed model software such as PROC MIXED and then PROC GLIMMIX was necessary and inevitable.

8.2 THE NEW STANDARD

In contrast to the world of the 1990s, the term linear model now means the generalized linear mixed model. What was called the linear model in 1990 is just a special case and one that, as we have seen here, is inadequate for the typical demands of modern research. Standard statistical practice now assumes generalized linear mixed models as the basic tool of analysis. As we have seen in Chapter 7, the GLMM should find more and more use as a planning tool for the design of research experiments, as well as for their analysis.

We have also learned that generalized linear mixed models are a good deal more complex than what passed for standard statistical practice in the past. Several obvious questions present themselves.

- Is the gain worth the added complexity?
- What are the consequences of remaining in the past and not using GLMM methodology?
- Assuming the gain does justify the added complexity and the consequences of remaining in the past are unacceptable, how does the agricultural and natural resources sciences community adapt to the rather drastic changes in standard statistical operating procedure that have occurred in the past decade?

The first two questions can be addressed together. Several of the examples, particularly in Chapters 4 and 5, have compared the results of analyses with GLMMs with results that would have been obtained using pre-GLMM methods. Generalized linear mixed model analyses consistently have two advantages, namely, efficiency and accuracy. Chapter 7, focusing on planning and design, reinforced these advantages from a somewhat different perspective.

Efficiency concerns the power of statistical tests and the precision of statistical estimates. Power characterizes the ability of a statistical test to identify a treatment effect, if indeed it exists, to an agronomically relevant degree in a scientifically defensible manner. Precision refers to the ability of a statistical procedure to estimate a treatment mean or the magnitude of a treatment effect with an acceptable margin of error; that is, precision refers to the width of a confidence interval. In the presence of random effects, spatially or serially correlated data, or non-normal data, generalized linear mixed models typically have greater power and precision for the same sample size than competing procedures. For example, consider the comparison of the probability of a desired outcome for two treatments in a randomized block design. The seed germination example presented in Chapters 2 and 5 showed that the normal approximation to the binomial and the arcsine square root transformation—two standard pre-GLMM practices—both yielded less precise estimates and less powerful tests than the generalized linear mixed model. This example illustrated the conditional versus marginal model issue that pervades analyses with non-normal data and complex designs. It is the statistical version of "you can run, but you can't hide." Non-normal data are inevitable in modern research. Split plot and repeated measures designs are a fact of life in agricultural research. Whenever these two elements are present, the conditional–marginal model issue exists. If one uses pre-GLMM methods, one is using the less efficient marginal distribution approach, like it or not.

Chapter 7 makes it clear that while the principles underlying the design of experiments have not changed, the way these principles play out and the design choices they suggest are different for non-normal data, often in unexpected ways. A design that conventional wisdom and experience suggests should be perfectly adequate may be catastrophically inadequate if the primary response variable is non-normal. This is because most of the accumulated wisdom about design in the agricultural sciences has been acquired via normally distributed data and pre-GLMM linear model based theory. Generalized linear mixed model theory can be applied to assist with design choices. Several examples in Chapter 7 suggest that, in many cases, much smaller designs than those conventional wisdom would suggest are needed can be used without loss of power or control over type I error. In other cases, those same tools can identify designs that are grossly under-powered and inadequate for the stated objective, saving researchers much wasted effort. Collecting insufficient data still costs money and effort that achieve nothing if the experiment is badly conceived.

Efficiency is a particularly important issue now when public research universities and other research entities in the agricultural sciences face ongoing fiscal constraints, tight resources, and shrinking budgets that are unlikely to change in the foreseeable future. If generalized linear mixed model based methods can achieve higher quality information with the same amount of data or information of equal quality with less data as the examples shown here demonstrate, then they can and should be used.

In non-technical terms, accuracy means estimating what you think you are estimating. This occurs in two ways in the examples that were presented here. In the split plot and repeated measures examples, mixed model methods are impera-tive to determine the correct standard errors for the treatment effects of primary interest. Pre-mixed model procedures simply cannot do this. In the presence of non-normal data, the accuracy issues are exacerbated, as we saw in the split plot and repeated measures examples in Chapter 5.

The second way the accuracy issue occurs can be illustrated by returning to the binomial, two-treatment randomized block design. As we saw in the seed ger-mination example, the probability one estimates with non-GLMM methods (the normal approximation and the arcsine square root transformation) is not the prob-ability one thinks one is estimating. We saw that the conditional GLMM model expresses the probability in this experiment as it is universally understood, but only the conditional GLMM model actually estimates it. This is an extreme version of the accuracy issue.

As research grows in complexity and the penalty becomes increasingly severe for the kinds of inaccuracy demonstrated in the examples, especially as in Chapter 5, what passed for standard methodology a decade or two ago will become increasingly unacceptable.

8.3 THE CHALLENGE TO ADAPT

Now, how do we address the third and most difficult question? How does the agricultural and natural resources sciences community adapt to the rather dras-tic changes in standard statistical operating procedure that have occurred in the past decade? Clearly, the standard statistical methods for agricultural researchers' curriculum that have been the staple of courses for graduate students over the past several decades do not prepare them to implement generalized linear mixed model analyses. So, one challenge to both the agricultural and statistics faculty at universities is determining what these courses should look like. The subject mat-ter taught in these courses has changed relatively little in the past 50 years, but standard statistical practice has changed dramatically in the past decade. These courses must adapt. The question is not whether, but how?

The second challenge relates to the nature of collaboration. It is nearly a cli-ché to say that the best research in the future will be team-oriented research, not single investigator research. However, like most clichés, this one has a basis in truth. It is tied to an old joke that circulates among statistical scientists who work in agriculture, "There are three areas in which people believe that they can be experts without any formal training: law, medicine, and statistics." The complex-ity of generalized linear mixed models, their "you can run, but you can't hide" and "what you don't know can (and probably will) hurt you" aspects make it clear that researchers must come to terms with generalized linear mixed model issues and that there is great deal more to their implementation than was the case for ANOVA and regression.

As we have written in this book, our guiding belief has been that agricul-tural researchers can and should learn the basics of GLMM methodology and that many of these procedures can be implemented by non-statisticians. At the same time, we have also gained a new appreciation for the complexity of generalized linear mixed model theory. Acquiring expertise in GLMM theory and methods is a full-time job. Agricultural scientists should not expect to be self-sufficient in statistical design and analysis. Research at the level conducted today demands the collaboration of members of research teams as equals. Successful teams will have statistical scientists as fully participating members of those teams.